B. Waterhouse Hawkins

Artistic Anatomy of the Horse

B. Waterhouse Hawkins

Artistic Anatomy of the Horse

ISBN/EAN: 9783741180354

Manufactured in Europe, USA, Canada, Australia, Japa

Cover: Foto ©berggeist007 / pixelio.de

Manufactured and distributed by brebook publishing software
(www.brebook.com)

B. Waterhouse Hawkins

Artistic Anatomy of the Horse

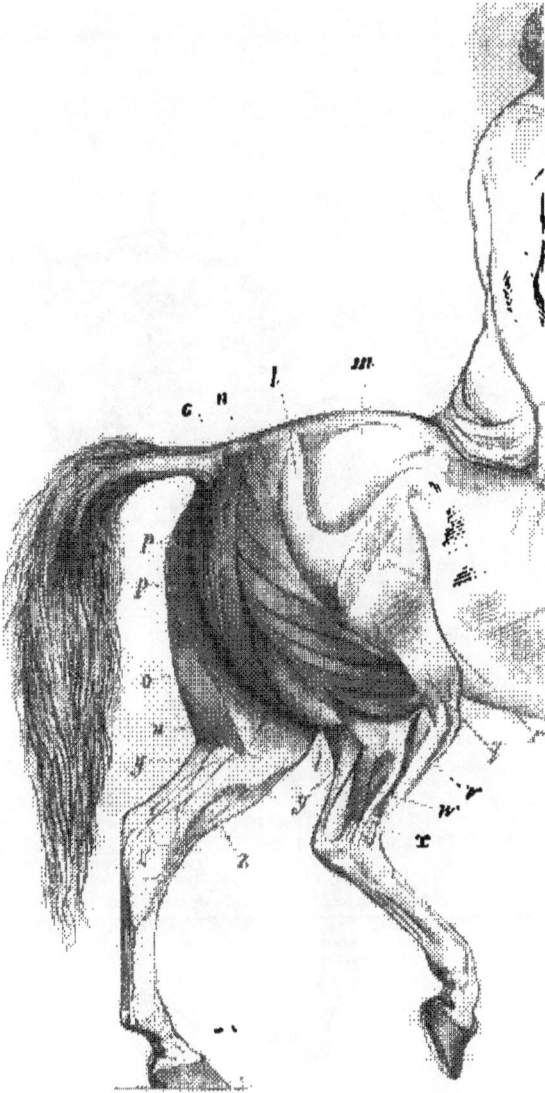

Pl.

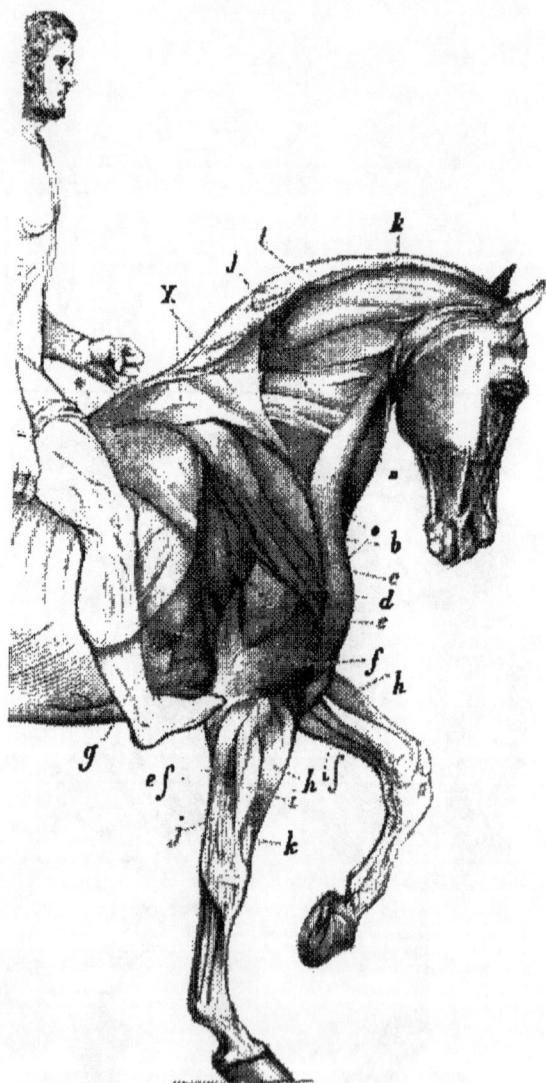

Tf. I.

THE

ARTISTIC ANATOMY

OF

THE HORSE.

BY

B. WATERHOUSE HAWKINS, F.L.S. F.G.S.

AUTHOR OF

" POPULAR COMPARATIVE ANATOMY," " ELEMENTS OF FORM,"
" COMPARATIVE VIEW OF THE HUMAN AND ANIMAL FRAME," AND
RESTORER OF THE EXTERNAL FORMS OF THE EXTINCT ANIMALS AT
THE CRYSTAL PALACE PARK, SYDENHAM.

With Twenty-four Illustrations drawn on Wood by the Author.

Ars probat artificem.

LONDON:

WINSOR AND NEWTON, 38, RATHBONE PLACE,

Manufacturing Artists' Colourmen, and Drawing Paper Stationers, by Appointment,
to Her Majesty, and to H.R.H. the Prince of Wales.

1865.

LIST OF PLATES.

PLATE.   PAGE.

1. Muscles of the Horse . . . . *frontispiece*

2. Skeletons of Horse and Man. 12

3.   Do   (*comparative view*). 20

4. Bones of the Head and Neck. 26

5. Muscles of the Head and Neck 31

6. Front view of Horse (*muscles*). . . . 41

7. Bones and Muscles of the Fore Legs. . . . 45

8. Bones of the Knee (*carpus*). 49

9. Bones of the Foot 54

10. Front limb of Horse and arm of Man. . . . 57

11. Bones of the Hock (*tarsus*). 64

12. Hind limb of Horse and leg of Man. . . . 67

13. Muscles of the Hind Quarters. 70

14. Back view of Horse (*muscles*). 73

THE
ARTISTIC ANATOMY
OF
THE HORSE.

BEFORE proceeding to the Anatomy of the Horse, it may be desirable to glance at some of the varieties of form which have been represented in the works of the earlier masters, either of painting or sculpture; and to observe how far they may be of value for imitation by the art-student of the present day. Such varieties are not very numerous, and those most frequently introduced present distinctions often peculiar to the times in which they were painted.

The external characteristics of the horse are strongly marked, and, as will be shown hereafter, are, to a large extent, due to the proportions and prominence of particular parts of the bony frame or skeleton. Constant, however, as are these general characters, the figure of the horse is as varied as the purposes for which man employs him, and to which in obedience to the demands of fashion, and the requirements of industry or

sport, a careful attention to scientific breeding has been
able to adapt it. Although such variations in external
form are particularly marked at the present day, when
the horse is so universally employed, and when, as a
rule, each description of work is performed by an animal
having qualities especially adapted for it, such was not
the case in earlier times. Horses were then chiefly required for warlike purposes, and there was little need of
any great diversity in their form or character.

In early historic times the horse was rarely, if ever,
used for agricultural purposes. Oxen were then, as in
many countries, they are now, the only animals employed
for draught in the laborious operations of husbandry.
In the East and in the warmer parts of Europe, the
horse was generally of a lighter and smaller type than
those in the West and North. It was well-fitted by its
activity for the purposes of light draught and was in
constant use in the chariot—speed and good action being
among the most essential qualities of the horses thus
employed.

We find examples of them in the Nineveh Sculptures,
which give such evidently faithful representations of
the light and graceful horses then in use, that we need
only copy them with care to ensure entire accuracy.
The Greeks have also left us complete fac-similes of the
horses of their time. The sculptures known as the
" Elgin Marbles" in the British Museum are perfect
studies from nature, and from these the art-student or
amateur, after he has acquired a thorough knowledge of

the skeleton, may obtain all that can be learned of the
general side view of the horse.

If we turn now to the works of the great masters of
the sixteenth and seventeenth centuries we find a particular type of horse generally represented. This is
what, not very long ago, artists delighted in calling the
"historical" horse—a large-boned, heavy animal in appearance, intermediate between a dray-horse and a
carriage-horse, such as we see in the pictures of Velas-
quez, Rubens, Vandyke, Wouvermans and others. These
were the horses of the period—the great Flemish and the
Norman horses, whose bone and substance were essential
to the fashion of the times which required them as
chargers to carry into battle a weight of armour, accoutre-
ments, and weapons of the heaviest description, or its
mimic-field, the tournament, to move with slow and
stately pace under the massive trappings of their knightly
riders. The horse in the well-known equestrian statue
of Charles I. at Charing Cross is deservedly regarded as
an excellent representation of the once famous breed of
Flemish horses, now, however, replaced by others more
suited to the requirements of modern warfare.

The light dragoon with his small active horse must,
however, be considered as quite a recent feature in what
are now commonly spoken of as battle pieces. The typical
Norman horse may still be seen on the Continent, in the
few places where that almost extinct conveyance, the
diligence drags on a lingering existence; and the great
Flemish horse with his high bones, massive muscles and

flowing mane still holds his place in his native country.
But it must not be forgotten that these animals so commonly represented in the old paintings belonged to a
national type, which, in course of time, and in accordance
with the prevailing fashion, gradually extended into other
lands. It became the model in various schools of art
because it was the particular breed in general use, and
was familiar to the people in such scenes as painters
loved to represent. It need hardly be mentioned, however, that this historical horse can rarely have a place in
the works of modern artists, particularly those of the
English school. Fresh breeds have been introduced,
and English types now furnish the student or amateur
with abundant models of the forms best suited to the
varied subjects of his pencil.

For a due comprehension of the shape and action of
the horse it is essential that the art-student should
possess a thorough acquaintance with the arrangement
and position of the bones, and their relation to one another. The osseous framework or skeleton must therefore be attentively studied as the only foundation upon
which can be based any hope of power to represent the
horse, either by drawing, painting, or modelling. The
importance of this first step cannot be too strongly in-
sisted upon, for without a knowledge of the bones, the
precise situation of the muscles cannot be determined,
or their action on their limbs properly understood.
Having mastered the details of the skeleton, the muscles
will then demand attention: and when once a general

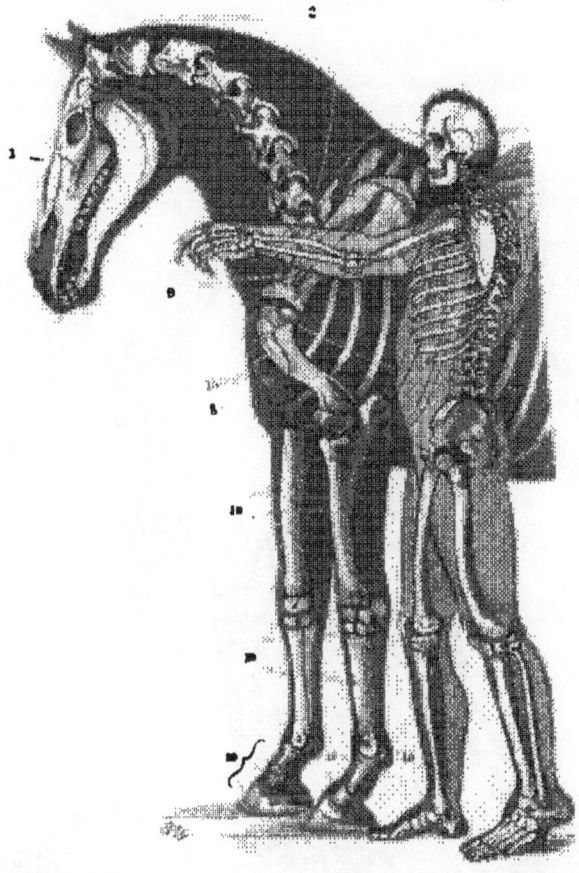

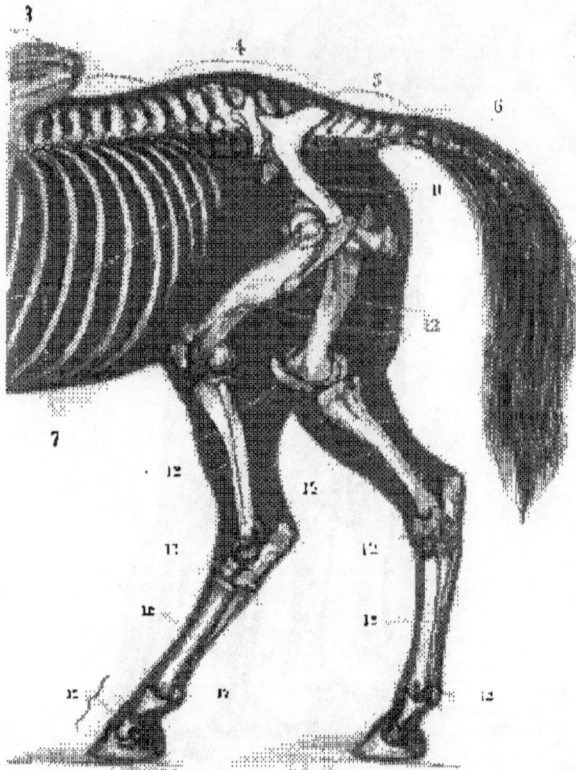

knowledge of them has been acquired, the eye accustomed to habits of accurate observation will readily discern in the living horse, the true appearance of those parts of the muscular system which are most prominently brought into view, and on whose alteration of form, when called into action, the minute but important changes of outline in the various parts of the animal principally depend.

PLATE II.

THE SKELETON OF THE HORSE.

1. Cranium.   See Plate IV.
2. Cervical vertebræ.   See Plate IV.
3. Dorsal vertebræ.
4. Lumbar vertebræ.
5. Sacral vertebræ.
6. Caudal vertebræ.
7. Ribs.
8. Sternum.
9. Scapula.   See Plate X.
10. Front limbs.   See Plate VII.
11. Pelvis.   See Plate XII.
12. Hind limbs.

THE SKELETON.

The series of bones comprised in the skeleton may be

conveniently divided into two groups, the first compre-
hending the head, neck and backbone with the ribs,
sternum and haunch bones—in other words, the
vertebral column and all those bones which are in
immediate connection with it, together constituting
the frame-work of the trunk, the second group comprising the bones of the limbs or legs, divided into a
double series of somewhat complicated articulations or
joints, necessary for the safety of the animal under the
manifold shocks and strains to which its varied and often
violent action continually exposes it.

In Plate II. is given a view of the entire skeleton of
the horse, showing the various bones in their natural
position and relation to one another. The head may,
for descriptive purposes, be divided into two parts—the
skull and the face; each having its own particular bones,
whose relative size varies in some slight degree in the
different breeds, and considerably affects the intelligent
expression more or less to be observed in the face of
every horse. These bones will be more particularly described in the chapter on the bones of the head and neck.

The bones composing the vertebral column are divided
into five groups.

The *cervical* vertebræ (Pl. II, 2. Pl. IV, fig. 2), or those
of the neck, extending from the head to the ribs, are
seven in number in the horse, as in all other mammalia.
Eighteen are given to the back, and are called *dorsal*
(Pl. II, 3); these are the only ones bearing the ribs.
The third group contains six vertebræ, the *lumbar* (Pl. II,

4), or those of the loins, situated between those bearing the
ribs, and the haunch bones. It should be borne in mind,
however, that the number of ribs is sometimes found to
exceed that stated above; nineteen, and, occasionally,
twenty ribs are found in the horse; but in such cases
there is no actual increase in the number of bones in
the vertebral column. The *dorsal* and *lumbar* vertebræ
together are always twenty-four, so that if one or two
ribs above the normal number are present, thereby in-
creasing the contents of the *dorsal* series, the *lumbar*
vertebræ are proportionately reduced. The fourth (Pl. II,
5), the *sacral*, includes five bones which are anchylosed
or united together into one mass, and, thus joined act
as a kind of wedge or keystone to the arch formed by
the approximation, at this point, of the haunch bones.
Great strength and solidity are required here, as the
united bones of the haunch, or *pelvic arch*, as they are
called by anatomists, are the great pivots on which
the hinder limbs turn, and by which they are enabled
to throw forward the whole weight of the animal. The
remaining vertebræ are those of the *caudal* (Pl. II, 6),
usually fifteen ; they are, however, exceedingly subject to
variations to the extent of two or three above or below
the number above mentioned.

To recapitulate, the normal contents of each series of
vertebræ will stand thus :

Cervical, 7 ; Dorsal, 18 ; Lumbar, 6 ; Sacral, 5 ; Caudal, 15 ; total, 51.

The form of these bones varies considerably in the

different parts of the vertebral column. It will be unnecessary, however, to describe them very minutely, as, except
in the case of those of the back, their shape does not
conspicuously affect that of the animal. The most prominent feature in each dorsal vertebra is the strong
spinous process or projection on its upper surface. These
processes are largely developed on the anterior portion
of the dorsal series, and produce the elevation or prominence above the shoulder, commonly called the *withers.*
They are of considerable importance to all long-necked
quadrupeds, from their affording a large surface for the
attachment of the great ligament which supports the
head and neck. All together they form the ridge of the
back. On each side of the dorsal vertebræ *transverse*
processes are situated, articulating with the ribs; and
other smaller oblique projections serving to unite and
fit one vertebra to that adjoining. The spinal column
has considerable flexibility, as well as very great strength;
these essential qualities being due to pads of cartilage
interposed between the several bones, and firmly
united to them. Besides these there are ligaments running along the broad under surface of the vertebræ;
others again between the transverse processes, and
similar strengthening ties uniting the upright projections or spinous processes, the whole mass forming a marvel of strength, lightness and flexibility.
The ribs (Pl. II, 7), eighteen in number, are jointed to
the transverse processes of the vertebræ, and curve with
some variations in their outline and direction, down

towards the sternum or breast-bone, to which the first
seven or eight of them called the *true* ribs, tho number
sometimes varying, are attached by their extremities, which
to provide tho elasticity necessary for the expansion of
the chest, are composed of cartilage. Tho remaining ribs
are termed *false* ribs, as they have no individual connection with tho breast-bone; they arc, however, united
together by cartilages, each on its own side, and this car-
tilaginous union ultimately terminates in the sternum;
so that the whole of tho ribs are enabled to expand or
act in uniformity. The *sternum* (Pl. II, 8), in the young
horse consists of six bones, which become united into a
single piece in the full-grown animal. The front of this
bone is convex and sharply keeled, its upper extremity
projecting so as to be easily observed in tho living horse.
This is known as the " point of the breast," and its place
will be casily ascertained when it is remembered that the
lowest part of the collar just covers it.

The haunch or pelvis (Pl. II, 11. Pl. XII, 6), is in
reality made up of six bones—three on each side, the
whole firmly united into one. Of these the *ilium* is the
most important, and is strongly secured to the *sacral*
vertobræ, which we have already noticed as forming the
keystone of the pelvic arch. Lateral prolongations of the
ilium produce the prominences so conspicuous just above,
and in front of the hind quarters in every horse. Tho
ischium or hip-bone is a backward continuation of the
ilium, and bears a considerable tuberosity which projects
on each side a little below the tail. The *pubis*, apparently

a single bone, is connected with those already mentioned, and forms an inverted arch with them below Pl. XII, d.

The bones of the limbs next claim our attention.

The natural attitude of the horse being that of a quadruped supported on the *extremities* of its four limbs, and with its body in a horizontal position, there is a greater apparent difference between its skeleton and that of man than really exists, as will be evident after a very slight examination and comparison of the two series of bones composing them.

In Pl. III. we have given a comparative view of the two skeletons in as nearly as possible the same attitude. It will be observed that besides the greater length of jaws and neck in the horse, (although the number and arrangement of the bones in these parts are the same in both skeletons) the principal differences consist in the form of the extremities and the uses to which they are applied. Man rests on the entire length of the foot, and his hands and fingers are constructed for grasping. The horse, on the contrary, is supported on the extreme points of its toes and fingers, reduced on each limb to a single digit, and protected by the nail becoming modified into a hoof. The long-established phraseology of horsemen, brought into use by the necessity for distinguishing the different parts of fore-legs and hind-legs, and the strange confusion resulting from the introduction of new names, and the misapplication of old ones, render a due comprehension of the nature and relation of these

PLATE III.

limbs almost impossible to those who have given no attention to the skeleton of the horse.

Many of the bones in each skeleton are known by the same names, but some parts of the limbs in the horse have been strangely miscalled. We have thought it desirable, therefore, to give in parallel columns, the names of those bones and joints which, although exactly corresponding in man and the horse, are spoken of under different titles in the ordinary description of the two skeletons. Our references to the various woodcuts will show the true relation and correspondence of the several bones, but as we shall have to speak specially of the horse, it will perhaps be desirable to employ the terms generally used in connection with that animal.

NAMES COMMONLY APPLIED TO CORRESPONDING BONES IN MAN AND THE HORSE.

Front Limbs.

Man.		Horse.
Arm (humerus)	=	Lower bone of shoulder.
Fore-arm	=	Arm.
Wrist (carpus)	=	Knee.
Hand (metacarpus)	=	Leg, cannon and splints.
Knuckles	=	Fetlock.
Finger	=	Pastern and foot.

Hind Limbs.

Thigh (femur)	=	Upper bone of thigh.
Knee	=	Stifle joint.

C

Man.		Horse.
Leg	=	Thigh.
Ancle *(tarsus)*	=	Hock.
Heel	=	Point of Hock.
Foot *(metatarsus)*	=	Leg.
Ball of Foot	=	Fetlock.
Toe	=	Pastern and foot.

The fore-leg or front limb is united to the body of the horse by means of the shoulder which is here said to be composed of two bones, both covered in and hidden by numerous over-lying muscles. The upper bone or shoulder blade has the usual flattened and long triangular shape, and is strengthened by a ridge or crest dividing it longitudinally into two somewhat unequal portions. The shoulder-blade or *scapula* rests on the ribs, the short side or base of the triangle being placed just below the withers, and its point directed downwards and forwards, nearly on a level with the top of the breast bone. The shoulder-blade has no osseous connection or articulation with the body of the horse, but is united to it solely by muscles, which will be spoken of more in detail in our chapter on the shoulder. The clavicles or collar bones, so well known in man and a few quadrupeds, do not exist in the horse. The lower bone of the shoulder, as it is commonly called, corresponds to the *humerus* or upper bone of the human arm. It is a short, thick and somewhat twisted bone articulating by a rounded head with the *glenoid* or cup-shaped cavity at the point of the shoulder-blade. Its lower extremity, which is directed

backwards, terminates in two condyles, receiving between them the head of the principal upper bone of the
leg. This portion of the fore-leg, commonly called the
arm (*fore-arm,* human,) is composed of two bones, a
long one in front termed the *radius* which extends to
the knee, and a short one behind called the *ulna.* The
latter bone has a long projection above and behind the
upper joint, and forms the point of the elbow to which
some powerful muscles are attached for extending the
arm. It rapidly diminishes in size towards its lower
extremity, and terminates in a point before it reaches
the knee. In old horses these two bones of the arm become firmly united into one.

The knee is a complicated joint uniting the arm to
the shank or leg, and is composed of six small bones
interposed between the upper and lower portions of the
fore-leg. We shall have occasion to describe the struc-
ture of this important joint at greater length in a subse-
quent chapter. Below the knee are the metacarpal
bones or those of the leg. They are three in number,
the cannon, and two splint bones behind. They represent the bones of the human hand, those between the
wrist and the fingers. The remaining bones of the fore-
leg are the upper and lower pasterns, and the coffin bone
surrounded by the hoof or nail, together forming a single
stout finger—the only one developed.

In the hind-leg we find a very similar arrangement of the bones. We have already spoken of the
pelvic arch, made up of the several bones of the haunch.

c 2

At a point on the outer surface of the pelvis, and at the junction of the three component bones on each side, a deep cup-shaped cavity called the *acetabulum* is formed to receive the round head of the true thigh bone or *femur* (Pl. XII. *e*). Great strain is thrown on this joint, it is therefore well protected by the bony cup or *acetabulum*, to whose centre the head of the femur is further secured by an exceedingly strong ligament. The *femur* or true thigh bone is so much concealed by the large muscles of the hind quarters that its true relations, or even its existence may not be recognized in the living horse. This circumstance has led to the confusion of names into which horsemen have fallen when speaking of the different parts of the hind leg.

The lower extremity of the femur is united to the bones of the true leg (*tibia*) by the " stifle joint," which also includes the *patella* or knee cap (Pl. XII. *f.*), this joint corresponding to the knee in human anatomy. The bones of the leg (' thigh,' of horsemen) are the *tibia* and *fibula* (Pl. XII. *g. h.*) articulating below with the numerous small bones of the ancle. The "hock" (Pl. XII. *i.*) is formed by a number of small bones, one of them having an elongated lever-like form with its free extremity directed upwards. This is the *os calcis* or bone of the heel. Into this bone the tendons of several powerful muscles are inserted, and a great deal of the springing power of the horse, as well as in other jumping animals, is due to the position and action of this part of the hinder limb.

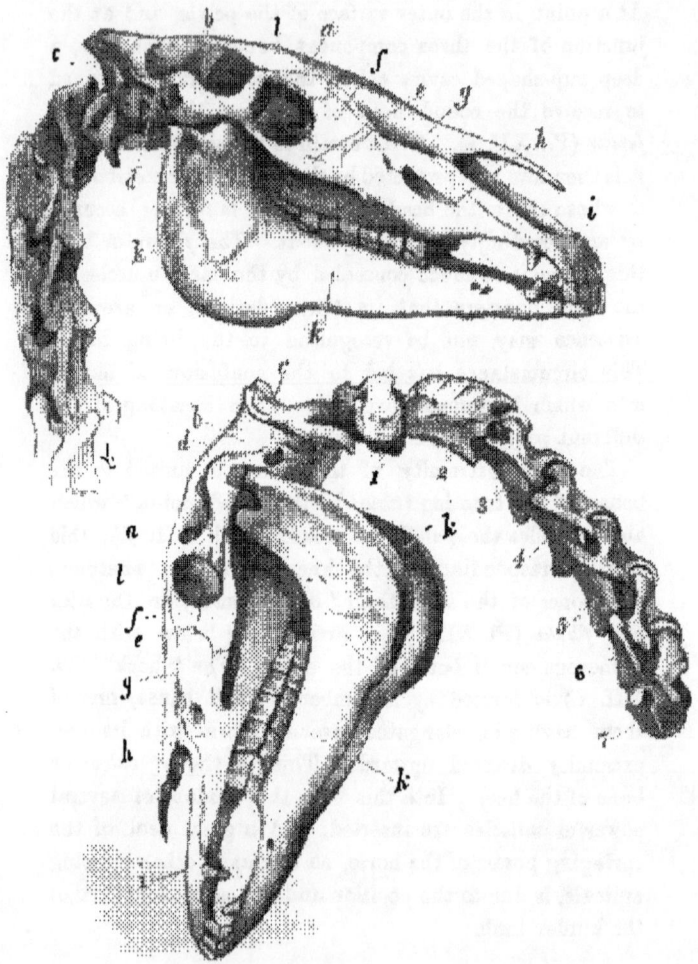

PLATE IV.

The remaining bones of the hind leg agree generally with those of the corresponding portions of the anterior limb.

PLATE IV.

FIGS. 1 & 2.

BONES OF THE HEAD AND NECK.

a. Frontal.
b. Parietal.
c. Occipital.
d. Temporal.
e. Malar.
f. Lacrymal.
g. Nasal.
h. Superior maxillary.
i. Pre-maxillary.
k. Inferior maxillaries or lower jaw.
l. Orbit.
1. Atlas.
2. Dentata.
3. Third.
4. Fourth. } Cervical vertebræ.
5. Fifth.
6. Sixth.
7. Seventh.

The bones of the head may be divided into two groups, those of the cranium and of the face. The cranial bones include all those which cover or enclose the brain. They are for the most part arranged in pairs, one on each side

of the mesial line of the skull, but may conveniently be
spoken of as single bones. ·

The *frontal*, or bone of the forehead (*a*) forms the
broad flat surface between the eyes, and extends with
a narrowing outline towards the top of the head. The
frontal occupies the widest part of the head. Considerable
difference in the width of this bone may be noticed in
various horses, and it will generally be found that the
broad and ample forehead is a mark of high breeding
and superior intelligence in the animal, as is often sufficiently indicated by the expression of the face. The
parietal (*b*) extends backward from the frontal to the
poll. It has a ridge or crest of great strength and
hardness along the upper surface, from which the bone
slopes down like a roof on each side, covering the brain,
which it is mainly concerned in protecting.

Immediately behind the *parietal*, and covering the
entire back of the head, is the *occipital* (*c*), a bone whose
position exposes it to greater strain than any of the other
component parts of the skull are liable. The *occipital*
has to support the whole weight of the head, which is
articulated by two rounded protuberances or *condyles* at
the base of this bone to the *atlas* or first vertebra of the
neck. On the outer sides of the *occipital*, and beyond
the *condyles*, are two styliform processes or pointed projections for the attachment of some of the muscles of the
neck which assist in supporting the head.

The *temporal* bone (*d*) unites above with the *parietal*, and
behind with the *occipital*. It contains the internal parts

of the ear, and has a depression or hollow beneath for
the articulation of the lower jaw. Anteriorly, this bone
joins the extremity of the *frontal*, and continuing forward
unites with the *malar* or cheek-bone (*e*), making up the
zygomatic arch, and forming the greatest part of the
orbit, which is completed by the *lacrymal* (*f*), a small
facial bone at the inner corner of the eye. Immediately
before the frontal is the *nasal* bone (*g*), one of the principal bones of the face, and covering the delicate mem-
brane of the nose. The *superior maxillary* (*h*), is a
large bone occupying the side of the face. It carries all
the molar teeth or grinders and the tusk of the upper
jaw. The nippers or incisor teeth are inserted in the
pre-maxillary (*i*), which uniting with the two bones last
mentioned completes the frame-work of the nose. The lower
jaw consists of two bones only, the *inferior maxillaries* (*k*).
These are rounded at the hinder extremity of the jaw
and terminate in two processes directed upwards.

The terminal projection or *condyloid process* articulates,
with the *temporal* bone at the base of the *zygomatic arch*,
and forms the hinge on which the whole lower jaw moves.
The second process, termed the *coronoid*, passes under the
arch, and receives the lower end of the large *temporal* muscle
which arises from the parietal bone, and is principally
concerned in moving the jaw in the act of mastication. There are also two small bones, in the lower
part of the cranium, under the *parietal*, the *sphe-
noid*, and *ethmoid*; they serve to connect the principal bones of the skull, but as they are not visible

externally, they do not need description for artistic
purposes. The bones of the neck, as we have already
mentioned, are seven in number. The *atlas*, which
articulates with tho skull, is a ring-shaped bone with
broad lateral projections, but without any other prominent characters. It has great freedom of motion on the
second bone, or *dentata*, and on the peculiar articulation
of these two vertebræ tho power of turning tho head
mainly depends. The remaining five bones of the neck
closely resemble one another; they have various small
processes for the insertion of muscles and ligaments, and
their form will be sufficiently understood by an examination of Plate IV.

PLATE V.

FIGS. 1 & 2.

MUSCLES OF THE HEAD AND NECK.

Head.

a. Masseter.
b. Temporalis.
c. Orbicularis.
d. Levator.
e. Orbicularis oris.
f. Dilator naris lateralis.
g. Zygomaticus.
h. Nasalis labii superioris.
i. Depressor labii inferioris.

PLATE V.

Neck.

j. Complexus major.
k. Splenius.
l. Levator anguli scapulæ.
m. Hyoideus.
n. Sterno-maxillaria.
o. Levator humeri or deltoides.

The muscles of the head are not very numerous, and those requiring the most attention will be found in the immediate neighbourhood of the mouth and nostrils.

The largest superficial muscle is the *masseter*, (Pl. V, fig. 1, 2 *a*). This forms the cheek of the horse, and extends along a ridge by the side of the head, below the eye, to the rounded posterior angle of the lower jaw which has a roughened surface for its more secure attachment. Its action is to close the mouth. The *temporal* muscle (*b*) also assists in this office. It arises from the medial ridge of the parietal bone, clothing its roof-like walls, and is inserted within the zygomatic arch to the coronoid process of the lower jaw-bone. The dimpling which may be observed, during mastication, above the eye of the horse is produced by the action of this muscle in alternately raising and depressing the under jaw. The *orbicularis* (*c*) is a circular muscle surrounding the eye and closing the eyelids. Above the eye, and directed inwards and upwards is a small *levator* muscle (*d*) which passes over the *orbicularis* and raises the upper eyelid.

The muscles of the ear are not very conspicuous.
Three of them may bo shortly noticed. The first proceeding from the base of the ear extends a short distance
forward and turns it in that direction; the second, behind the ear, directs it inward and backward, and
the third descends as a narrow strip at the back of the
cheek to incline the ear outward.

The frontal and nasal bones have no prominently
perceptible muscular covering; the difference in the shape
of these parts in various horses being entirely due to the
variation in the relative size and proportion of the particular bones.

Of the muscles of the lips and nose we may first men-
tion the *orbicularis* (*e*), one of the most important of them.
It entirely surrounds the mouth, and by its action the
lips are pushed out or closed. This muscle is brought
into play whenever the lips are required to seize or hold
anything between them. The *dilator naris lateralis* (*f*) is
a pyramidal muscle covering the whole exterior of
the nostril and having its origin close to the anterior
point of the masseter. It is the great *side* dilator
of the nostril, and also raises the upper lip. The *zygomaticus* (*g*) draws back the corner of the mouth
whence it may be traced upward, outside the masseter to its origin on the zygomatic arch. The
buccinator, a muscle on the inside of the mouth and
cheek, and consequently scarcely visible externally, has
the same office as the preceding.

The *nasalis labii superioris* (*h*) extends from a depres-

sion in front of the eye towards the angle of the mouth,
a short distance above which it divides into two parts,
the side dilator of the nostril (*f*) passing between them.
One of these portions is continued straight to the
corner of the mouth which it raises; the other part ex-
pands under the side dilator, and assists it in the office
of dilating the nostril. It also helps to lift the upper
lip.

The under lip is drawn back by the *depressor labii
inferioris* (*i*), a narrow muscle which is inserted into the
lip below the angle of the mouth, and passing along the
side of the jaw, disappears under the masseter.

Independently of the muscles for supporting the head
and neck there is a very beautiful and simple arrangement
by which those parts are kept in an easy and natural
position when the horse is at rest. This consists of a
very strong and elastic ligament called the *ligamentum
nuchæ*. It takes its origin from the back of the occipital
bone to which it is attached immediately below the
crest. At first it is in the form of a stout round cord.
It passes over the *atlas*, or first joint of the neck, to
allow full freedom of motion to the head, and is strongly
adherent to the *dentata*, on which the principal strain
from the weight of the head is thrown; it then proceeds
backward to its termination on the elevated spinous processes of the first dorsal vertebræ. The *withers* as these
elevated parts are called have thus an important office—
that of supporting the weight of the entire head and neck
when in their ordinary position. But provision must

also be made for lowering and raising the head, and for
these purposes there are special muscles. The first to
be noticed is the *complexus major* (Fig. 1. *j*). It arises
from the transverse processes of the four or five first
dorsal vertebræ, and also from the five lower bones of the
neck; the fibres from these two points uniting to form
one large muscle which diminishing in size in the direction of the head terminates in a tendon inserted into the
occipital bone. This muscle makes up the principal portion of the lower part of the neck. Immediately above
this is the *splenius* (*k*) specially employed in raising the
head. It arises from the entire length of the *ligamentum
nuchæ* and is directly inserted into all the bones of the
neck, except the first, with which, however, and the
temporal bone of the head, it has a separate and less
distinct connection. To the form and development of
the splenius, the beauty of the neck of the horse is mainly due. It is here the greatest thickness is found; and
from being sometimes overloaded with cellular substance
or fat, an appearance of clumsiness may be produced.
The thick crest and massive neck of the entire horse are
to a large extent due to the abundant development of
this muscle; and the student or amateur will do well
to acquire a thorough knowledge of its form, which, in
every condition and breed of the horse so largely con-
tributes to give a character to the neck.

Behind the splenius, and extending along the superior
margin of the neck is the *levator anguli scapulæ* (*l*). It
is inserted into the back of the head, and attached to the

first four bones of the neck as well as to the great ligament, then descends to the shoulder where it is not visible externally. It has a reciprocal action on the neck and shoulder according to whichever is the fixed point at the time.

Of the muscles in front of the neck we may first direct attention to the *hyoideus* (Plate V, fig. 2, *m*). Its upper extremity is always conspicuous immediately below the head at its junction with the neck. It is attached to the hyoid bone of the tongue, which it retracts, and descends along the front of the neck to the shoulder, but is covered in the greatest part of its length by other muscles, and is only visible for a short distance below the head. Outside this muscle, and partly covering it, is the *sterno maxillaris* (Plate V. *n*), the principal depressor of the head. It arises from the upper end of the sternum or point of the breast, covers the lower front of the neck, then proceeding upwards by the side of the hyoideus, is inserted by a flat tendon into the posterior angle of the lower jaw. It is not a very large muscle, for, when those supporting the head and neck are relaxed, but little force is required to pull the head down.

Beyond the sterno-maxillaris, and extending from the back of the head and upper part of the neck, along the front of the shoulder, to the top of the fore-leg, is the *levator humeri* or *deltoides* (Plate V. *o*), a long and very important muscle, having, in fact, a double function to perform. When the head is kept up by its own proper muscles, it becomes a fixed point from which the *levator*

38   THE ARTISTIC ANATOMY

humeri is enabled to raise the shoulder. This is pro-
bably its principal office. Its action, however, can also
be reversed, and with the shoulder for a fixed point, the
head can be depressed, a small slip of the muscle being
carried forward to the point of the sternum to pull the
head in that direction.

NOTE.—It must be borne in mind that, with very few
exceptions, the muscles are all arranged in pairs, sometimes though rarely in contact, and that in speaking of
them in the singular number, unless otherwise stated, we
are referring to their position and function on each side
of the animal.

BONES AND MUSCLES OF THE SHOULDER.

The shoulder blade or *scapula* (Plate II, 9, Plate X.
fig. 1, *a*) consists of a single bone, and connects the
fore-leg with the trunk, corresponding in its relation to
that of the haunch bone to the hind-leg. There is, however, this important difference between them, the haunch
bones are anchylosed or united to the sacral portion of
the back-bone in order to provide a firm point from
which those powerful levers, the hind-legs can act; the
shoulder, on the contrary, has to receive a violent shock
from the weight of all the front part of the animal sud-
denly falling on the fore-legs. The shoulder has therefore only a muscular attachment to the trunk; and by
this arrangement no jar is received by the spine, and any

injury to the important viscera of the chest is rendered unlikely.

The shoulder-blade is of a long triangular form, with its apex directed downwards, nearly on a level with the point of the breast, and its somewhat rounded base resting on the ribs immediately below the withers. It is divided externally into two portions by a ridge or crest running nearly the length of the blade, and a little on one side of its mesial line. This ridge of bone gives additional firmness to the shoulder-blade, and affords a surface for the attachment of some very important muscles. At the lower extremity of the shoulder-blade is a cup-shaped hollow, called the *glenoid cavity*, with which the rounded head of the bone (*humerus*) of the shoulder articulates. Above this joint, on the anterior edge of the scapula, is the *acromion process*, to which in man and some few quadrupeds the clavicle or collar bone is united. This bone, however, is not found in the horse, or in other animals which have but little power of lateral motion in the front limbs.

Following the custom of horsemen, and adopting their nomenclature for the bones of the horse, we shall next speak of the " lower bone of the shoulder" the *humerus* (Plate II, 10, Plate X, fig. 1, *b*), in every respect corresponding with that part of the human arm which extends from the shoulder to the elbow, but which, in the horse, is so hidden by the muscles as not to be externally visible as a distinct bone of the front limb. The lower bone of the shoulder is short and strong; it articulates by a rounded head with the glenoid cavity of the scapula, and has considerable freedom of motion. Its direction is backwards, and at almost a right angle with the shoulder-blade. It has several large protuberances at the upper end of the bone, and to which are attached the principal muscles for moving it. The lower extremity terminates in two condyles or heads between which the superior end of the arm-bone is received.

PLATE VI.

MUSCLES OF THE SHOULDER AND BACK.

z. Trapezius.
a. Pectoralis minor.
b. Antea spinatus.
c. Postea spinatus.
d. Teres minor.
e. Anconœus longus.
f. Anconœus externus.
g. Serratus major.
*. Latissimus dorsi.
p.m. Pectoralis major. See Plate XIV.

Of the muscles of the shoulder we may first notice the *trapezius* (Plates I, VI, XIV). It rises from the ligament of the neck and the principal bones of the withers, and terminates in a pointed shape on a prominent part of the ridge of the shoulder-blade. Its office is to raise and support the shoulder, assisting the *serratus major* (Plates

PLATE VI.

I, VI, g), a very important muscle, but hardly visible externally, as it is principally situated between the shoulder-blade and the ribs of the horse, forming the main connection between them.

The *antea spinatus* (Plates I, VI, XIV, b), taking its name from its situation, occupies the outer surface of the scapula on the front side of the spine or ridge of that bone. It proceeds to the lower bone of the shoulder, and, dividing into two parts, is inserted into the two prominences in front of it, extending the bone forwards. The *postea spinatus* (Plates I, VI, XIV, c) is situated on the other side of the spine of the shoulder-blade, and is inserted into the upper and outer head of the bone, drawing it outward and raising it. Behind the *postea spinatus* is a small muscle called the *teres minor*, (Plates I, VI, XIV, d), or little pectoral; it draws the shoulder forward towards the breast. The *pectoralis major* (Plato XIV, p.m.) is conspicuous inside the arm at its junction with the body. It is an important muscle, and pulls the whole fore-leg inwards, keeping it on a line with the body, and ensuring an even and regular action of the limb. On the outside of the shoulder, and readily seen in the living horse when in motion, are two muscles, which, arising from the lower bone of the shoulder, are inserted into the point of the elbow. They are called the *anconæus longus* (Plates I, VI, XIV, e), and the *anconæus externus* (Plates I, VI, XIV, f). Their office is to straighten and extend the arm, in other words, to bring the front limb into a perpendicular position, and as nearly as possible

D

in a line with the *humerus*, or, as we have called it, the
"lower bone of the shoulder." The muscles which bend
the arm upwards are not visible externally, but are almost entirely covered by those of the shoulder.

The muscles of the back do not require any lengthened
notice. The *latissimus dorsi* (Plates I, VI, XIV) is the
most important; it covers the whole back, extending from
the shoulder to the haunch, and is strongly attached to
the processes of the vertebræ, and the ribs. This muscle is the principal one employed in raising the fore or
hind quarters in the act of rearing or kicking. That
part of it which comes nearest to the surface is generally
covered by an ordinary saddle, but no portion of this
muscle is at any time very distinctly visible.

PLATE VII.

BONES AND MUSCLES OF THE FRONT LIMBS.

FIG. 1.—*Bones.*

A. Radius.
B. Ulna, point of.
c. Knee (*carpus*).
D. Cannon or Shank.
E. Splints.
F. Sesamoids (behind Fetlock).
G. Upper and Lower Pasterns.
H. Coffin Bone.
I. Navicular.

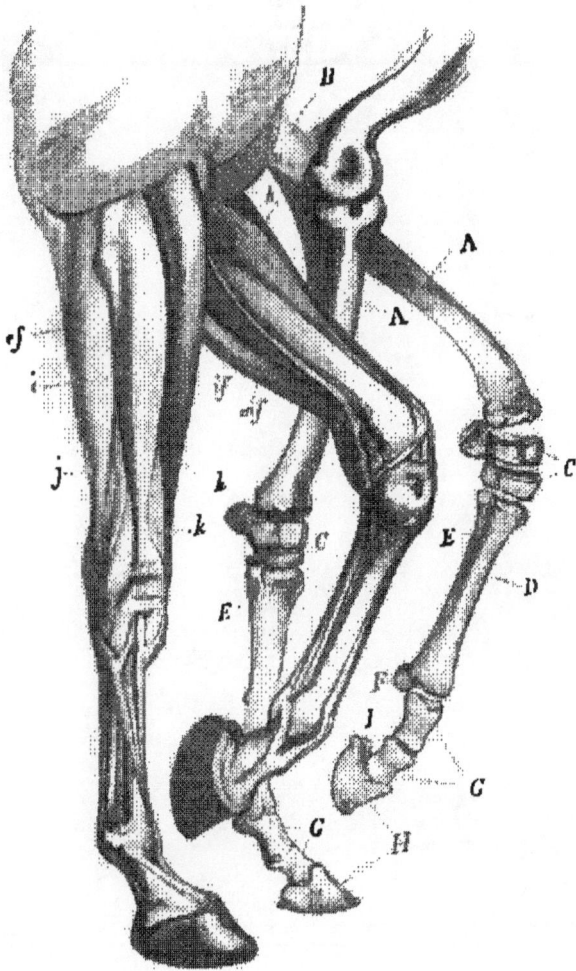

PLATE VII.

FIG. 2.—*Muscles.*

h. Extensor carpi radialis.
i. Extensor digitorum longior.
j. Extensor digitorum brevior.
k. Abductor pollicis longus.
ef. External Flexor.
mf. Middle Flexor.
if. Internal Flexor.

The upper portion of the fore-leg, or, as it is commonly called in the horse, the arm, (*fore-arm*, human) extending from the elbow to the knee (*carpus*), consists of two bones, the *radius* and the *ulna.* The *radius* (Fig. 1. A) is the more important of the two, and in the young horse is the great support of the leg. It is the long front bone, is nearly straight, and receives into depressions on its upper end the two heads of the inferior extremity of the lower bone of the shoulder. The other end of the *radius* fits on to the upper layer of the bones of the knee (*carpus*). The *ulna* (fig. 1. B.) is situated behind, and to some extent above, the *radius*, there being a considerable projection of the former received between the heads of the lower bone of the shoulder, and called the elbow. This forms a powerful lever into which are inserted the muscles for extending the arm as already noticed in our account of the muscles of the shoulder. The ulna is continued downwards, gradually diminishes in size, and terminates in a point behind the middle of the radius. These two bones of the arm are at first distinct and

separate, but before many years have past the carti-
laginous and ligamentous connection between them
becomes ossified, and the two bones are firmly united
into one.

The knee, (Fig. 1, c) corresponding to the human wrist
(carpus), is a part of the fore-leg to which the attention
of the artist should be particularly directed, as its form
is always 'a characteristic and prominent feature in the
outline of the horse, and one to which, like the hand in
the drawing of the human figure, severe scrutiny is
likely to be applied. The knee is a complicated joint,
that is, it is composed of numerous small bones interposed between the lower end of the radius and the upper
extremity of the shank or cannon bone.

The position and action of this joint render it pecu-
liarly liable to external injury and violent jars or strains ;
it is therefore so made up that any shock it may receive
will be distributed over a number of distinct bones, each
protected by a covering of cartilage, and resting on a
kind of semi-fluid cushion, the whole being strongly
united together by ligaments.

PLATE VIII.

BONES OF THE KNEE (carpus)

FIG. 1. *Left leg, outer side.* FIG. 2. *Front view.*

a. Radius.
b. Pisiforme.

PLATE VIII.

FIG. 1.

FIG. 2.

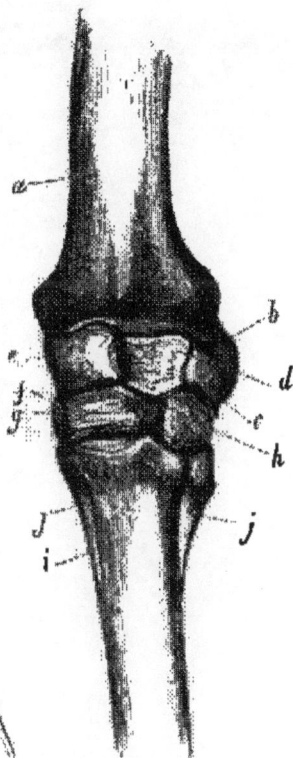

c. Cunieforme.
d. Lunare.
e. Scaphoides.
f. Trapezoides.
g. Magnum.
h. Unciforme.
i. Cannon. } Metacarpals.
j. Splint. }

The true carpal bones are seven in number, six of them being being placed in two rows, each containing three bones, in front of the joint, and the seventh, the *pisiforme* (Plate VIII, figs. 1-2, *b*), by some persons called the *trapezium*, being situated behind them, forming the point of insertion for some of the muscles of the arm, and otherwise aiding in the protection of the tendons running down behind the leg. By reference to Plate VIII, the shape and arrangement of the bones of the knee will be readily understood. Besides the advantage of distributing a shock over several distinct parts, there is another object to be gained by the interposition of these small bones. The bending of the leg at this point can be carried so far that a very wide opening between the bones of the arm and the shank would necessarily be the result, and this would take place at a part extremely liable to external injury. By the presence and arrangement of the interposed carpal bones, however, this wide opening is replaced by three narrow ones, which are well protected from all ordinary dangers by being covered with a capsular ligament, extending from

tho radius above, to the shank bone below them. A large
flat knee has always been considered a valuable point in
a horse, and from what we have shown of the action of
this joint, the advantages of its possessing a considerable
extent of surface will be sufficiently evident.

Between the knee and the fetlock are three bones, the
cannon or *shank*, and two *splint* bones, the whole making
up what is called the leg (*metacarpus*, human). The
cannon or *shank* bone (Plate VII, fig. 1, D), articulates
at its upper extremity with the lower row of the bones
of the knee, and at the other end with the upper pastern
at the fetlock joint. It is the principal bone of this
portion of the leg, and is almost entirely devoid of any
muscular covering, those parts of it which are not hidden
by tendons being only protected by the skin. This bone
is nearly straight, rounded in front, and flattened or
slightly concave behind. The *splint* bones (Fig. 1, E) are
situated behind the cannon and a little on each side of
it. They also articulate with the lower bones of the
knee, and throughout their length are united by cartilage and ligaments to the cannon bone. The name
given to these bones well describes their character, they
are "splints"—slender pieces attached to the cannon to
strengthen it, and diminishing to a point before they
reach the fetlock joint (Plate IX, c). Behind this
joint are two small supplementary bones termed .
sesamoids (Plate IX, b; Plate VII, fig. 1, F.); they serve
to protect the back of the joint and some important
ligaments passing over it.

PLATE IX.

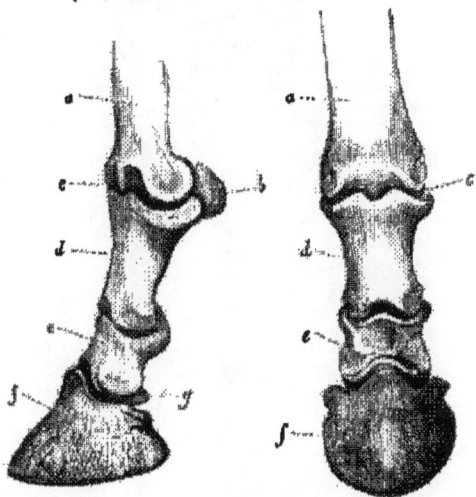

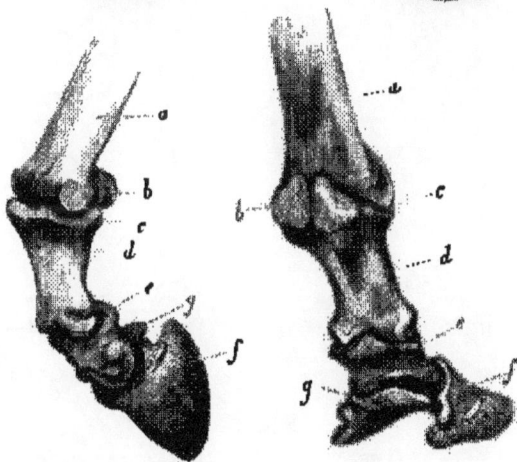

PLATE IX.

Figs. 1—4.

Bones of Foot.

a. Cannon or Shank.
b. Sesamoids.
c. Fetlock joint.
d. Upper pastern.
e. Lower pastern.
f. Coffin bone.
g. Navicular bone.

The two next bones in descending to the foot are the *upper and lower pasterns* (Plate IX, d, e, Plate VII, fig 1, a) ; these have considerable motion one on the other to allow the foot to be bent back. The toe is formed by the *coffin* bone, (Plate IX, f, Plate VII, fig. 1, B) which is surrounded and covered in by the horny hoof, so that its form is never visible externally. For all artistic purposes, the shape of the hoof need only be considered. Another small bone called the *navicular* (Plate IX, g, Plate VII, fig. 1, I.) is found behind, and partly within, the junction of the coffin and lower pastern, and like the former bone is enclosed by the hoof.

PLATE X.

COMPARATIVE VIEW OF THE BONES IN THE FRONT LIMBS OF
MAN AND THE HORSE.

a. Scapula.
b. Humerus.
c. Olecranon or elbow.
d. Radios.
e. Ulna.
f. Carpus.
g. Metacarpus.
h. Digit.
1. Phalanx.
2. Do.
3. Do.

In order to render more intelligible the relation of the
several bones of the fore-leg in the horse to those of the
human arm and hand, there is given in Plate X, a comparative view of those parts of the two skeletons, by
which it will be seen that any differences existing between
them are due to alterations in the shape or proportions,
or in some cases, to the suppression or undevelopment of
particular bones, but not to any departure from the
general plan on which both skeletons are constructed.
A general agreement in form, although not in proportions, will be noticed in the shoulder blades (a) and the
humeri (b) or first bones of the actual limbs; the same

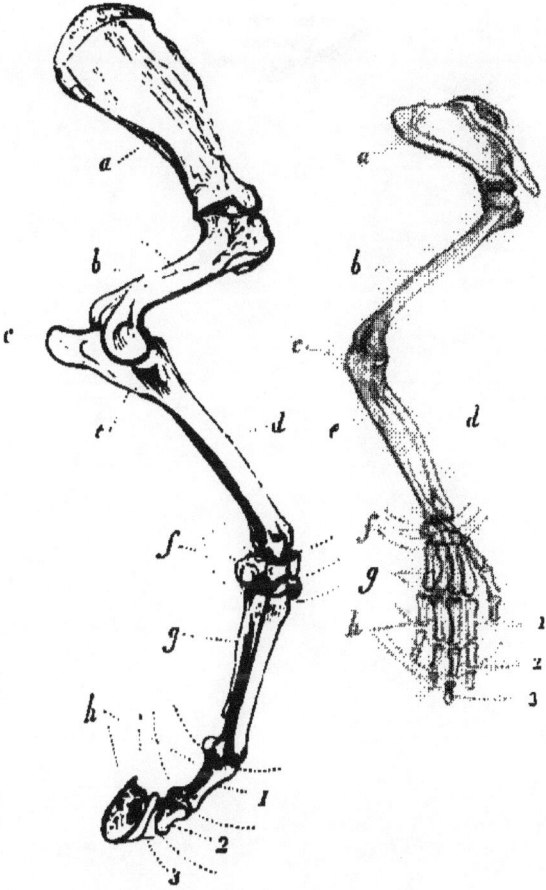

PLATE X.

may be observed in the next bones of the series, but with
a slight modification. The *radius* (*d*) is the same as in
both skeletons, but the human *ulna* (*e*) is completely
developed, and terminates at the wrist (*carpus*) which
gains additional power of motion by its articulation with
this second bone. There is also a general correspondence
in the carpal bones (*f*). In the metacarpals, (*g*), however, we find an important distinction. The five bones
bearing this name in the human hand are reduced to
three in the front limb of the horse, where they are considerably increased in relative size, and are known as
the *cannon* and *splint* bones. Only one of the fingers is
developed in the horse, the middle digit, corresponding
to the middle finger (*h*) of the human hand, and the
three bones composing it are in the horse known as the
two *pasterns* (1-2), and the *coffin* bone (3). The hoof, as
we have before mentioned, is only a modified form of
nail. In the accompanying woodcut those bones of the
human hand which have their homologues in the horse
will be found strongly indicated.

The muscles of the fore-leg may be divided into two
groups, those in front, and on the outer side of the limb ;
and those which may be seen from behind, and on the
inside.

The fleshy portions of all these muscles are placed
above the knee (*carpus*), and only their tendinous prolongations are continued to the several bones of the
lower part of the leg and foot.

The principal muscle in front of the so-called " arm"

of the horse is the *extensor carpi radialis*, (Plates I, VI,
VII, XIV, *h*).

It arises from the lower part of the lower bone of the
shoulder, and descends in front of the arm to the knee
where it becomes entirely tendinous. It then passes
over the knee, under a band of ligament which crosses
that joint, and is finally inserted in front of the cannon
bone. The action of this muscle is to strengthen the
lower part of the leg. Next to this muscle are those
whose office it is to extend the foot, the *extensor digitorum longior* and *extensor digitorum brevior*, Plates I,
VI, VII, *i. j.*) Their origin is much the same as that of
the extensor of the leg, but the tendons pass by the side
of the knee, under the capsular ligament of that joint,
down in front of the leg and of the fetlock joint to be
inserted into the pasterns and coffin bone. The first of
these muscles is conspicuous on the outside of the arm,
but the second is in a great measure hidden by its com-
panion.

A small oblique muscle the *abductor pollicis longus*
(Plates I, VII, *k*), appears from under those last mentioned, and obliquely crosses the knee; it assists the
others in extending the leg.

On the outside of, and rather behind the arm is the
most external of the muscles which bend the leg, the
flexor carpi ulnaris, or external flexor (Plates I, VI, VII,
XIV, *ef*). It arises from the outer head of the lower
bone of the shoulder, and descends towards the knee,
the tendon dividing into two parts, one of them being

inserted into the *pisiforme* (Plate VII, fig. 1), the seventh bone of the knee, and conspicuous behind that joint, the other going to the outer splint bone. On the inside of the leg, and behind it is the middle flexor (Plates VII, XIV, *mf*) springing from the inner hand of the lower bone of the shoulder and terminating, like the outer flexor, at the *pisiforme*. These two muscles are the principal flexors of the leg, and are assisted in their office by the internal flexor (Plates I, VII, XIV, *if*) which having much the same origin as the others, is inserted into the inner splint bone. The muscles which bend back the foot are deeply seated, and covered by those we have just described. The tendons in which they terminate will be seen in the several figures in Plates I, VI, VII, XIV.

BONES OF THE HIND LIMB.

It will be unnecessary to give any detailed account of the bones of the haunch, as they have been sufficiently described in our general sketch of the skeleton. We shall therefore now proceed to point out the characters of the bones of which the hind leg is composed.

Beginning at the upper extremity of the limb, the first bone to be noticed, is the *femur* or true thigh (Pl. XII. *e*); and here we must direct the student's attention to the list of the names applied to corresponding bones in the skeletons of man and horse, (see page 13)

that he may become familiar with the true relation of
the several parts of the hind leg, so commonly misnamed
when speaking of the horse. We shall, as before,
use the forms most familiar to the equestrian, ex-
plaining them as may appear desirable for the due
comprehension of the subject. This bone (*femur*) is so
entirely hidden by various muscles of the haunch as
to be unrecognised and unnamed by those persons who
are not acquainted with anatomy. We shall speak of
it as the "upper bone of the thigh," a term that may
be easily remembered by those who apply the name of
"thigh" to the next lower bone of the series.

The "upper bone of the thigh" (*femur*) is exceedingly
strong and stout. It is short for its bulk, which is
further augmented by several large projections or *troch-*
anters placed longitudinally for the attachment of some
important muscles. The upper extremity of the femur
has a distinct rounded head on the inner side, fitting
into and articulating with the *acetabulum* or bony cup
formed at the junction of the three pelvic bones. The
lower end of the bone bears two prominences which fit
into corresponding depressions in the next bone, and
in front of which is placed the *patella* or knee-cap (Pl.
XII. *f*) together making up the "stifle joint" of horsemen, or, more strictly speaking, the actual "knee" of anatomists. The "thigh" (*leg*, human) consists of two
bones, the *tibia* (Pl. XII. *g*) and the *fibula* (Pl. XII. *h*).
The tibia extends from the stifle joint, which it helps
to form, to the "hock" (*ancle*, human). The fibula is

PLATE XI.

FIG. 1.   FIG. 2.

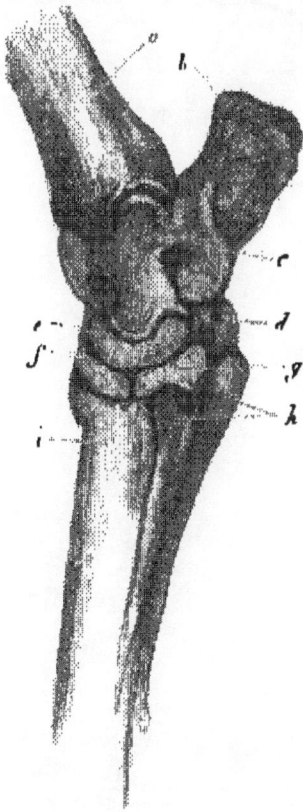 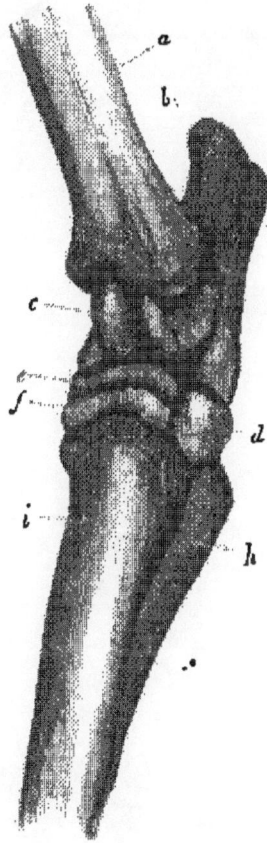

placed behind on the outer side of it, extending from
its upper extremity to about one third of its length.
It is attached to the larger bone by cartilage, and
agrees in general character with the small bone or ulna
in the fore-leg.

PLATE XI.

BONES OF THE HOCK, (*tarsus*).

FIG. 1. *Back view, inner side.* FIG. 2. *Front view, outer side.*

a. Tibia.
b. Os calcis.
c. Astragalus.
d. Cuboides.
e. Naviculare.
f. Outer cunieforme.
g. Middle cunieforme.
h. Splint.
i. Cannon or Shank.

NOTE.—As the great toe is unrepresented in the horse, the
inner cunieforme is not developed.

The *hock* (Pl. XI) is an important and somewhat complicated joint. It corresponds to the ancle and heel in
man, although, in the horse, it is at some distance from
the ground. Like the knee of the horse (*carpus*) the
hock (*tarsus*) consists of several small bones interposed
between the long ones of the lower part of the limb.

They are six in number, and of various shapes, for a
knowledge of which we must refer the student to Plate
XI. which gives a front and inner side view of the joint
with the several bones in their natural position. We
may, however, direct attention to the projecting bone
at the back of the joint. This bone, the *os calcis* or heel
bone forms what is called the "point of the hock." It
acts as a lever to straighten the leg, and is moved by
the *tendo Achilles* and other tendons arising from the
muscles which spring from the upper part of the limb.
It is considerably developed in all fast moving animals,
an increase in the length of the lever adding considerably
to the force of the spring.

PLATE XII.

COMPARATIVE VIEW OF THE BONES OF THE PELVIS AND HIND

LIMBS OF MAN AND THE HORSE.

a. Sacrum.
b. Ilium.
c. Ischium. } Pelvis.
d. Pubis.
e. Femur. .
f. Patella.
g. Tibia.
h. Fibula.
i. Tarsus.
j. Metatarsus.

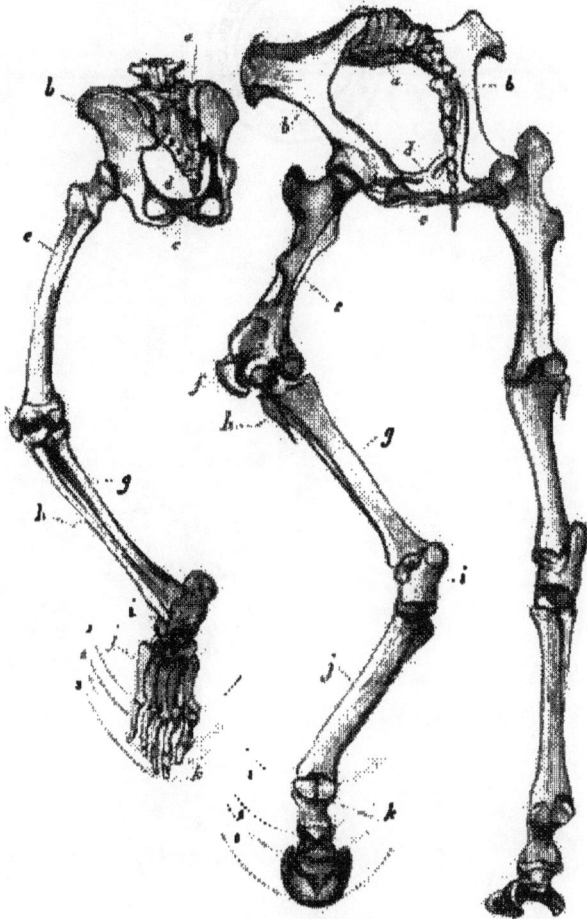

PLATE XII.

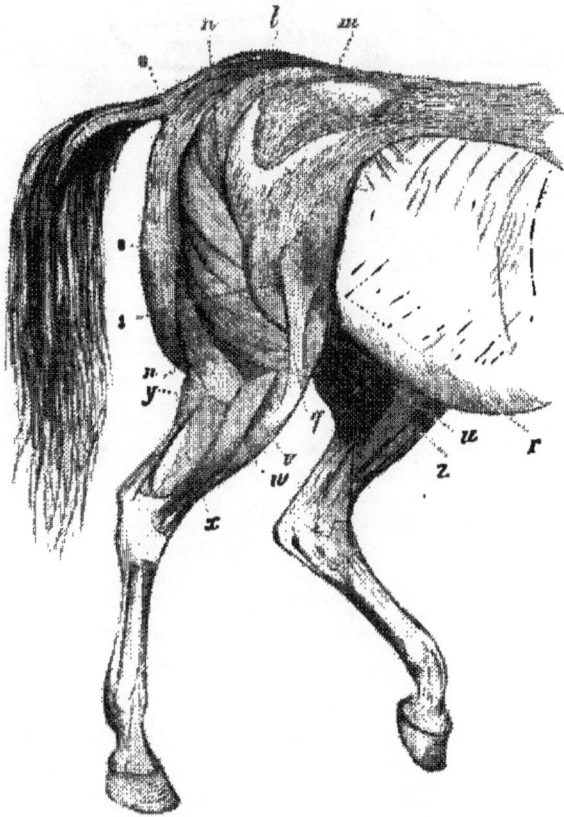

PLATE XIII.

k. Digit.
   1. Phalanx.
   2. Do.
   3. Do.

The remaining bones of the hind-leg do not require any lengthened description, as they agree generally with those in the lower part of the fore-leg. The "leg" (*metatarsus*, Pl. XII. *j*) is composed of the shank and two splint bones, the former uniting at the fetlock joint with the upper pastern, which is followed by the other bones of the toe, as in the front limb.

In Plate XII. we have given a comparative view of the hind limbs of man and the horse, by which the true nature and relations of the several bones may be readily understood, and to which our observations on a similar comparison of the front limbs are generally applicable.

PLATES XIII AND XIV.

MUSCLES OF THE HIND QUARTERS.

   l. Glutæus externus.
   m. Glutæus medius.
   n. Triceps femoris.
   o. Biceps.
   p. Semi-membranosus, Plate XIV.
   q. Musculus fasciæ latæ.
   r. Rectus.
   s. Vastus externus.

u. Gracilis.

v. Extensor pedis.

w. Peronæus.

x. Flexor pedis.

y. Gastrocnemi.

z. Flexor metatarsi.

Under this heading we shall include all the muscles which are concerned in, and connected with, the motion of the hind limbs.

The muscles of the hind quarters are for the most part strongly marked, and the situation of the principal ones easily recognized.

Prominent on the front and outer part of the haunch is the *glutæus medius* (Pls. I, XIII, XIV, *m*). It arises from the processes of several of the vertebræ of the loins, and from the prominent parts of the ilium, terminating at its insertion in the great trochanter or projection on the upper bone of the thigh *(femur)*. It is a very important muscle, and acts with considerable power in raising and bringing forward the femur. It has been called the "kicking muscle."

The *glutæus externus*, (Pls. I, XIII, XIV, *l*) is a slender muscle attached to the *glutæus medius*, having a generally similar origin and function.

Among the most conspicuous muscles of the hind quarter, especially when the horse is in motion is the *triceps femoris* (Pls. I, XIII, XIV, *n*), or three-headed muscle of the thigh *(femur)*. Strictly speaking it is made up of three muscles, but as they are united

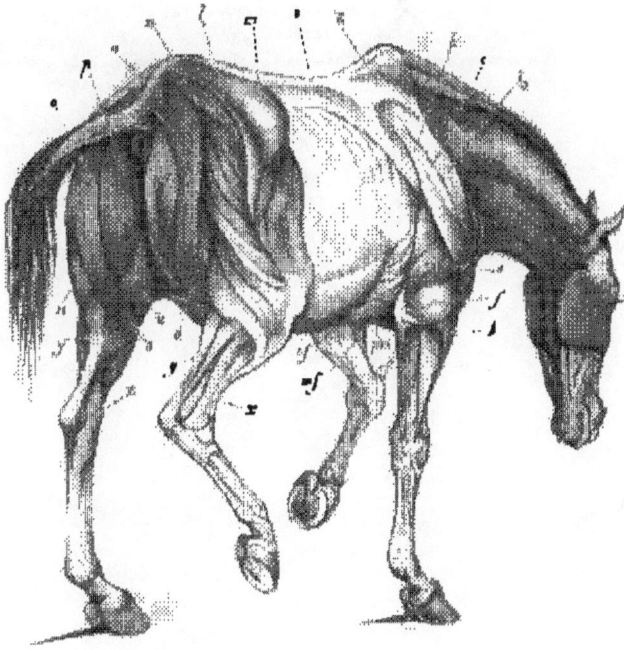

PLATE XIV.

and have a common action, it will be convenient
to speak of them as one. It takes its origin from
several of the bones of the spine, including some
at the root of the tail, and from various parts of the
haunch bones; it then curves downwards and forwards,
dividing into three heads which are inserted broadly into
the upper part of the lower bone of the thigh, behind
the "stifle joint" or true knee. Its action is evidently
to draw back the stifle joint, in other words, to straighten
the leg. It has therefore enormous power in impelling
the animal forward. The *glutæi* muscles bend the leg
preparatory to taking the spring, and the *triceps* acts in
opposition, forcing the leg straight, and consequently
lifting the body forwards. The posterior margin of this
muscle may be more or less distinctly observed, parallel
to the outline of the buttock, in all kinds of horses, but is
particularly evident in hunters and racers, where high
condition has resulted from the proper exercise of these
powerful springs of motion. Parallel with, and immediately
behind, the *triceps* is the *biceps* (Pls. I, XIII, XIV, *o*).
It springs from the sacrum and the first bones of the
tail, and descending to the inner side of the lower bone
of the thigh (*tibia*) forms the outer posterior border of
the haunch, and assists in straightening the leg. The
semi-membranosus (Pls. I, XIV, *p*), is also one of the
flexors of the leg; it forms the inner posterior border of
the haunch, and unites on the mesial line, under the
tail, with its fellow muscle of the other quarter.

On the outer front part of the haunch is the *musculus*

E

fasciæ latæ (Pls. I, XIII, *q*). It arises from the anterior
portion of the crest of the ilium and is enclosed between two layers of tendinous substance which disappears below the stifle. This peculiar muscle binds
down and secures the other muscles in front of the
haunch. The *rectus* (Pl. XIII, *r*) proceeds from the
ilium in front of the hip joint, and is inserted into
the patella or knee-cap. It forms the front edge of
the thigh. Behind the rectus, and also inserted into
the patella, is a large muscle called the *vastus externus*, (Pl. XIII, *s*) of which a part only can be seen
externally. These muscles are powerful extensors of the
thigh.

Descending inside the thigh is a narrow strip of
muscle terminating just below the stifle joint. This is
the *sartorius* or " tailor's muscles ;" it bends the leg (*tibia*)
and turns it inwards. It can hardly be seen. By the
side of this muscle, and to the rear of it, occupying the
principal surface of the inside of the thigh (*femur*), we
find the *gracilis* (Pls. I, XIII, *u*) inserted like the sartorius, into the upper part of the lower bone of the
thigh (*tibia*). Of the muscles which move the ¦lower
portion of the leg and the foot, the *extensor pedis* (Pls. I,
XIII, *v*) is the most important. It arises behind the
stifle, from the extremities of the two bones of the
thigh (*femur et tibia*) and descending to the hock, where
its tendon passes under a sheath confining it to its place
in front of that joint, continues its course to the foot,
and is inserted by a wide expansion into the front of the

coffin bone. The *peronæus* (Pls. I, XIII, *w*) follows much the same course as the last muscle, but takes a more lateral direction. It arises from the fibula, and the tendon passes on the outside of the hock, after which it descends to the foot with the tendon of the *extensor pedis*. These muscles lift the foot forwards. Between these muscles there is a small narrow one having the same function as the others, and whose tendon is visible just above the hock. The *flexor pedis* (Pls. I, XIII, XIV, *z*) is one of the principal muscles for bending the foot. It arises from the upper part of the tibia, and becoming tendinous before it reaches the hock, passes as a large round cord through a groove at the back of that joint; it then descends behind the shank bone to be inserted into the two pasterns. At the back of the "thigh" (*tibia*) the extremities of the *gastrocnemii* may be seen (Pls. I, XIII, XIV, *y*) with united tendons (*tendo achilles*) passing to the "point of the hock" (*os calcis*), where they are strongly inserted. There is some little difference between the development of the muscles whose tendons lead to the heel in man and the horse. In man, the artist will remember the *soleus* as forming the principal element in the great tendon of the heel. The *gastrocnemii* also contribute towards it. In the horse, however, these latter muscles take a more important share, and are aided by the *plantaris* which, in man, is extremely small. The *soleus*, on the other hand, is as little developed in the horse.

We may notice one muscle on the inside of the

"thigh" (*tibia*). In Pls. I, XIII, XIV, *z*, is the *flexor metatarsi* or bender of the "leg." It originates above the "stifle," on the upper bone of the thigh (*femur*), and is inserted into the shank and inner splint-bone. It lies just within the anterior margin of this "thigh" (*tibia*), and acts with considerable power in bending the hock, thereby raising the metatarsal bones. The metatarsus (Pl. XII, *j*) is entirely without muscular covering, its shape being solely due to the form of its component bones, and the position of the tendons and ligaments which pass over it in their descent to the pastern and foot.

THE END.

CATALOGUE

OF

MATERIALS FOR OIL PAINTING.

Index to Catalogue

OF

OIL PAINTING MATERIALS.

PAGES 25 TO 47.

	PAGE
Academy Boards	25
Bartholomew's Easel, for Sketching Oil	45
Brushes, Flat Hog Hair	43
,, Black Fitch	43
,, Round Hog Hair	43
,, Flat Sable	44
,, Round Sable	44
,, Flat Hog Hair Varnishing	44
,, Badger Softeners	44
Canvas, Fine Linen, prepared	35
,, on Frames	36
Chalks	47
Charcoal	47
Easels, Deal	46
,, Mahogany	46
,, ,, Rack	46
Grecian Painting	47
,, ,, Prepared Paper	47
,, ,, ,, Canvas	47

	PAGE
Mahl Sticks	46
Millboard Sketching Frame	46
Millboards, Prepared	37
Oil Colour Boxes	40
,, ,, Fitt	41
Oil Colours in Tubes	38
Oils	39
Palette Knives	47
Palettes	45
Panels, Superior Mahogany	37
Pencils, Red Sable Hair	44
,, Sable Writing	44
Powder Colours	42
Prepared Paper for Sketching	35
Slabs, Ground Glass	45
Tablets for Sketching in Oil	42
Tin and Japanned Ware for Oil Painting	46
Varnishes	39

CATALOGUE

OF

𝕸aterials required in 𝕺il 𝕻ainting.

FINE LINEN CANVAS.

PREPARED IN A SUPERIOR MANNER FOR OIL PAINTING, IN ROLLS OF
SIX YARDS LONG.

Warranted to keep any length of time without cracking.

					Canvas. s. d.	Roman. s. d.	Ticken. s. d.
½ or 27 inches wide	.	. per yard			2 0 ..	2 6 ..	3 0
⅞ or 30 "	.	"			2 3 ..	2 9 ..	3 3
36 & 38 "	.	"			2 6 ..	3 3 ..	3 9
3 feet 6 inches wide	.	"			3 3 ..	4 0 ..	4 6
3 " 9 "	.	"			4 6 ..	5 3 ..	6 0
4 " 6 "	.	"			5 3 ..	6 6 ..	7 6
5 " 2 "	.	"			7 3 ..	7 9 ..	8 6
6 " 2 "	.	"			8 3 ..	9 3 ..	10 0
7 " 3 "	.	"			12 0 ..		.. 14 0

PREPARED CANVAS.

IN PORTRAIT SIZES, WITHOUT FRAMES.

					Canvas. s. d.	Ticken. s. d.
Head size,	.	. 24 inches by 20	.	each	1 2 ..	1 9
Three-quarter,	.	30 " 25	.	"	1 9 ..	2 6
Kitcat,	. .	36 " 28	.	"	2 3 ..	3 3
Small half length,	.	3 ft. 9 in. by 2 ft. 10 in.	"		3 0 ..	4 0
Half length,	. .	4 - 2 " 3 - 4	"		4 6 ..	6 6
Bishop's half length,	4 - 8 " 3 - 8	"			7 0 ..	9 6
Whole length,	.	7 - 10 " 4 - 10	"		18 6 ..	22 6
Bishop's whole length,	8 - 10 " 5 - 10	"			25 0 ..	30 0
Hatchment Cloths,	.	3 - 9 " 3 - 9	.	. 4s. 6d. each.		
Ditto	.	4 - 6 " 4 - 6	.	. 7s. 0d. each.		

CANVAS PREPARED WITH PURE WHITE OR TINTED GROUNDS.

				Each
ACADEMY BOARDS, for Studies or Sketching, 24½in. by 18½				1s.
Ditto	Half Size	. .	18½ " 12¼	6d.

PREPARED PAPER FOR SKETCHING IN OIL,

Imperial size, 30 in. by 21 . . 1s. per sheet.

PREPARED CANVAS ON FRAMES.
ALL THE SIZES NAMED ARE KEPT IN STOCK.

Portrait Sizes.	CANVAS. Wedged Frames. s. d.		TICKEN. Wedged Frames. s. d.
24in. by 20, head size	2 3	..	2 9
27 „ 22, large ditto	2 6	..	3 3
30 „ 25, ¾ size	3 0	..	3 9
36 „ 28, Kitcat	4 0	..	5 0
44 „ 34, small half length	6 0	..	7 0
50 „ 40, half length	8 6	..	10 6
56 „ 44, Bishop's half length	11 6	..	14 0
7ft. 10 by 4ft. 10, whole length	30 0	..	34 0
8ft. 10 by 5ft. 10, Bishop's whole length	40 0	..	45 0
8in. by 6	0 8		0 9
10 „ 7	0 10		0 11
12 „ 10	1 0		1 1
14 „ 12	1 3		1 4
16 „ 14	1 6		1 8
20 „ 16	1 8		1 10
21 „ 17	1 10		2 0

Landscape Sizes.	CANVAS. Wedged Frames. s. d.		TICKEN. Wedged Frames. s. d.
9in. by 6	0 9	..	0 10
10 „ 7	0 10	..	0 11
12 „ 8	0 11	..	1 0
12 „ 9	1 0	..	1 1
13 „ 9	1 1	..	1 2
14 „ 10	1 2	..	1 4
15 „ 11	1 3	..	1 5
16 „ 12	1 4	..	1 6
17 „ 13	1 5	..	1 7
18 „ 12	1 5	..	1 7
19 „ 13	1 6	..	1 8
21 „ 14	1 8	..	1 10
22 „ 16	1 10	..	2 2
24 „ 18	2 0	..	2 4
27 „ 20	2 6	..	3 0
30 „ 20	2 9	..	3 6
36 „ 24	3 9	..	4 6

STRETCHERS COVERED WITH ROMAN CLOTH AT PROPORTIONATE
PRICES

Irregular and Large Sizes made to Order.

PREPARED MILLBOARDS.

PAINTED WITH OIL GROUNDS.

	s. d.		s. d.
6in. by 5 . . .	0 5	15in. by 11 . .	1 6
7 „ 5 . . .	0 6	15 „ 12 . .	1 7
8 „ 6 . . .	0 7	16 „ 11 . .	1 8
9 „ 6 . . .	0 7	16 „ 12 . .	1 9
9 „ 7 . . .	0 8	17 „ 13 . .	2 0
10 „ 7 . . .	0 8	17 „ 14 . .	2 2
10 „ 8 . . .	0 9	18 „ 12 . .	2 0
11 „ 8 . . .	0 9	18 „ 13 . .	2 2
11 „ 9 . . .	0 10	18 „ 14 . .	2 3
12 „ 8 . . .	0 10	19 „ 13 . .	2 3
12 „ 9 . . .	0 11	19 „ 14 . .	2 4
12 „ 10 . . .	0 11	20 „ 14 . .	2 6
13 „ 8 . . .	0 11	20 „ 16 . .	3 0
13 „ 9 . . .	1 0	21 „ 17 . .	3 3
13 „ 10 . . .	1 0	22 „ 18 . .	3 9
13 „ 11 . . .	1 1	23 „ 16 . .	3 9
14 „ 9 . . .	1 1	24 „ 18 . .	4 0
14 „ 10 . . .	1 2	24 „ 20 . .	4 6
14 „ 12 . . .	1 4	30 „ 25 . .	6 0

SUPERIOR MAHOGANY PANELS.

PREPARED ON THE FINEST WELL-SEASONED WOOD.

PAINTED WITH OIL GROUNDS.

	s. d.		s. d.
6in. by 6 . . .	1 0	16in. by 12 . .	3 9
9 „ 6 . . .	1 1	17 „ 12 . .	4 0
9 „ 7 . . .	1 2	17 „ 13 . .	4 3
10 „ 7 . . .	1 3	17 „ 14 . .	4 6
10 „ 8 . . .	1 4	18 „ 12 . .	4 3
11 „ 8 . . .	1 6	18 „ 13 . .	4 6
11 „ 9 . . .	1 9	18 „ 14 . .	4 9
12 „ 8 . . .	2 0	19 „ 13 . .	4 9
12 „ 9 . . .	2 2	19 „ 14 . .	5 3
12 „ 10 . . .	2 4	20 „ 14 . .	5 6
13 „ 8 . . .	2 2	20 „ 16 . .	6 6
13 „ 9 . . .	2 4	21 „ 17 . .	7 6
13 „ 10 . . .	2 6	22 „ 16 . .	7 6
13 „ 11 . . .	2 9	22 „ 18 . .	8 6
14 „ 9 . . .	2 6	23 „ 16 . .	8 0
14 „ 10 . . .	2 9	24 „ 18 . .	9 0
14 „ 12 . . .	3 3	24 „ 20 . .	10 0
15 „ 11 . . .	3 3	30 „ 25 . .	15 0
15 „ 12 . . .	3 6	36 „ 28 . .	25 0
16 „ 11 . . .	3 6		

OIL COLOURS IN PATENT COLLAPSIBLE TUBES.

Price 4d. per Tube.   Double Size, 8d.

YELLOWS.

Naples Yellow (Light)	Chrome Yellow
„ (Deep)	„ Deep
Yellow Ochre	„ Orange
Roman Ochre	Raw Sienna
Brown Ochre	Patent Yellow
Transparent Gold Ochre	King's Yellow
Orpiment	Italian Pink
Gamboge	Yellow Lake

REDS.

Burnt Sienna	Crimson Lake
Light Red	Scarlet Lake
Venetian Red	Purple Lake
Indian Red	Indian Lake
Burnt Roman Ochre	

BLUES.

Antwarp Blue	Permanent Blue
Prussian Blue	New Blue
Indigo	

BROWNS.

Vandyke Brown	Bone Brown
Cologne Earth	Cappah Brown
Raw Umber	Mummy
Burnt Umber	Verona Brown
Asphaltum	Brown Pink
Bitumen	

GREENS.

Emerald Green	Verdigris
Terra Verte	Olive Lake

BLACKS.

Blue Black	Ivory Black
Black Lead	Lamp Black

Oil Colours in Patent Collapsible Tubes, continued.

WHITES.

Flake White
Nottingham White
Blanc d'Argent
Permanent White

MEDIUMS.

Megilp
Copal Megilp
Pyne's Megilp
Sugar of Lead

EXTRA COLOURS.

	s.	d.			s.	d.
Burnt Lake	0	6	Mars Orange		1	0
Vermilion	0	6	Oxide of Chromium		1	0
Cobalt	1	0	Malachite Green		1	0
French Ultramarine	1	0	Mineral Grey		1	0
Indian Yellow	1	0				
Rose Madder	1	0	Lemon Yellow		1	6
Pink Madder	1	0	Cadmium Yellow		1	6
Rubens' Madder	1	0	Orange Vermilion		1	6
Brown Madder	1	0				
Mars Brown	1	0	Ultramarine Ash		2	6
Mars Scarlet	1	0	Carmine		2	6
Mars Violet	1	0	Burnt Carmine		2	6
Mars Yellow	1	0	Purple Madder		2	6

OILS, VARNISHES, &c.

	Phials. s. d.		Pint Bottles. s. d.
Mastic Varnish, double strength, for making Megilp	1 0	..	5 0
Mastic Varnish, for varnishing Pictures	1 0	..	4 0
Copal Varnish	1 0	..	5 0
Crystal ditto	1 0	..	5 0
White Spirit ditto	1 0	..	5 0
Brown ditto ditto	1 0	..	5 0
Lac Varnish	2 0	..	
Japan Gold Size	0 6	..	2 6
Nut Oil	0 8	..	3 0
Poppy ditto	0 6	..	2 6
Linseed ditto	0 4	..	1 6
Pale Drying ditto	0 6	..	2 0
Strong ditto	0 6	..	2 0
Fat Oil	0 6	..	2 0
Spirits of Turpentine	0 4	..	1 0
Asphaltum . . . . . per pot	0 6	..	3 0
Megilp . . . . . . "	0 6	..	
Copal Megilp . . . . "	0 6	..	

JAPANNED TIN OIL COLOUR BOXES,

FOR CONTAINING TUBE COLOURS, BRUSHES, OILS, AND THE MATERIALS REQUIRED IN OIL PAINTING.

	s.	d.
Japanned Tin Oil Colour Boxes, to hold 12 Tubes, Palette, Brushes, Oils, &c. Small size, 10in. by 6⅝, and 2 deep	8	0
Ditto, to hold 16 Tubes, &c. Size, 10in. by 6¾, and 2¼ deep .	10	0
Ditto, to hold 18 Tubes, &c. Size, 11¾in. by 7¾, and 2¾ deep	12	0
Ditto ditto, with double bottom and grooves, for containing Millboards and Wet Sketches, size 11¾in. by 7⅝, and 3¼ deep	15	0
Ditto, to hold 20 Tubes, &c. Size, 12¼in. by 8½, and 2¾ deep	12	6
Ditto ditto, with a double bottom and grooves, for containing Millboards and Wet Sketches, 12in. by 8. Size of Box, 12¾in. by 8½, and 3¼ deep	15	6
Ditto, to hold 27 Tubes, &c. Size, 12¾in. by 8½, and 3 deep .	15	0
Large size Tube Box, to hold 30 Tubes, &c., double bottom, for containing Millboards, 13in. by 9. Divisions for Japanned Bottles, &c., &c. Size of Box, 13½in. by 9½, and 3¼ deep	20	0
Flat Portable Tube Box, to hold 20 Tubes, Brushes, Oils, &c. Size, 12¾in. by 8½, and 1⅝ deep. A most convenient Box, particularly desirable when sketching from nature, or for Travelling	10	6
Ditto, the same Box, with double bottom, for containing Millboards, 12in. by 8. Size of the Box, 12¾in. by 8½, and 2 deep	12	6
Pocket Oil Sketching Box, to hold 12 Tubes, very portable. Size 13¼in. by 4¾, and 1¼ deep, with folding Mahogany Palette	8	6
Ditto ditto, to hold 16 Tubes, &c. Size, 9in. by 4⅝, and 2 deep, with folding Mahogany Palette	9	6
Ditto ditto, to hold 19 Tubes, &c. Size, 13¼in. by 4¾, and 1½ deep	7	6
Ditto ditto, to hold 24 Tubes, &c. Size, 13¼in. by 4¾, and 2 deep, with folding Mahogany Palette . . .	12	0

OPPOSITE ST. JAMES'S CHURCH.   41

JAPANNED TIN OIL COLOUR BOXES.

FITTED.

Containing Sets of Oil Colours in Tubes, an Assortment of Hog Hair and Sable Brushes, a Badger Softener, Oils, Varnish, Turpentine, Palette, Palette Knife, Portcrayon, Charcoal, Tin Dipper, &c., &c.

	£	s.	d.
Japanned Tin Oil Colour Box for 12 Tubes, fitted complete	1	1	0
Ditto, 16 Tubes, fitted complete	1	10	0
Ditto, 16 Tubes, with double bottom, containing Millboards	1	14	0
Ditto, 20 Tubes, fitted complete	2	4	0
Ditto, 20 Tubes, ditto with double bottom, containing Millboards	2	12	0
Ditto, 30 Tubes, Large Size, fitted with a full assortment of Brushes and extra colours in Tubes, double bottom, containing Millboards, 13in. by 9, very complete	4	10	0
Flat Portable Tube Box for 20 Tubes, fitted complete. Size of the Box, 12¾in. by 8½, and 1½ deep	1	10	0
Ditto, the same Box, with double bottom, containing Millboards	1	14	0
Pocket Oil Sketching Box, with folding Palette, fitted complete. A most convenient and portable Box when sketching from nature	1	0	0

TABLETS FOR SKETCHING IN OIL.

Made on the principle of the solid Sketch Blocks, being composed of a number of sheets of prepared paper for Sketching in Oil, compressed so as to form an apparently solid substance, but each sheet of which may be removed at pleasure, by passing a knife round the edge of the upper surface.

EACH BLOCK CONTAINS 32 SURFACES.

						s.	d.
8vo Imperial	10 inches by 7				each	5	0
4to	ditto	14	,,	10	,,	9	0
3mo	ditto	18	,,	10	,,	12	0
½	ditto	20	,,	14	,,	16	0

E 2

POWDER COLOURS.

GROUND WITH SPIRITS INTO IMPALPABLE POWDER, FOR OIL OR
WATER COLOUR PAINTING.

	£	s.	d.		£	s.	d.
Antwerp Blue, per oz.	0	1	4	Scarlet Lake, per oz.	0	6	0
Brown Madder „	0	6	0	Scarlet Vermilion „	0	1	0
Brown Pink „	0	1	4	Smalt „	0	4	0
Brun de Mars „	0	8	0	Ditto (Dumont's Blue)	0	18	0
Burnt Carmine „	0	16	0	Genuine Ultramarine	5	0	0
Cadmium Yellow „	0	12	0	Ditto „	4	0	0
Carmine „	0	16	0	Ditto „	3	0	0
Ditto „	0	12	0	Ultramarine Ashes	1	1	0
Ditto „	0	8	0	Ditto „	0	16	0
Cobalt Blue „	0	8	0	Ditto „	0	12	0
Ditto „	0	6	0	Verdigris „	0	1	4
Cremnitz White „	0	0	3	Vermilion „	0	0	6
Crimson Lake „	0	6	0	Yellow Lake „	0	1	4
Flake White „	0	0	3				
French Ultramarine,							
extra fine „	0	8	0				
Ditto „	0	6	0	Bone Brown „	0	1	0
Ditto „	0	4	0	Blue Black „	0	1	0
Green Oxide of				Brown Ochre „	0	1	0
Chromium „	0	5	0	Burnt Umber „	0	1	0
Indian Lake „	0	4	0	Burnt Sienna „	0	1	0
Indian Red „	0	1	4	Chrome Yellow „	0	1	0
Indian Yellow „	0	5	0	Cologne Earth „	0	1	0
Indigo „	0	1	4	Deep Chrome „	0	1	0
Italian Pink „	0	1	4	Emerald Green „	0	1	0
Jaune de Mars „	0	8	0	Ivory Black „	0	1	0
Lemon Yellow „	0	5	0	King's Yellow „	0	1	0
Madder Carmine „	0	16	0	Lamp Black „	0	1	0
Malachite Green „	0	4	0	Light Red „	0	1	0
Mineral Grey „	0	2	6	Naples Yellow „	0	1	0
Mummy „	0	1	4	Orpiment „	0	1	0
New Blue „	0	3	0	Orange Chrome „	0	1	0
Orange de Mars „	0	12	0	Patent Yellow „	0	1	0
Pink Madder Lake	0	6	0	Raw Sienna „	0	1	0
Prussian Blue „	0	1	4	Raw Umber „	0	1	0
Pure Scarlet „	0	5	0	Roman Ochre „	0	1	0
Purple Madder „	1	1	0	Terre Verte „	0	1	0
Purple Lake „	0	5	0	Vandyke Brown „	0	1	0
Rose Madder Lake	0	12	0	Venetian Red „	0	1	0
Rouge de Mars „	0	8	0	Yellow Ochre „	0	1	0

FLAT AND ROUND HOG HAIR BRUSHES IN TIN,
FOR OIL PAINTING.
POLISHED CEDAR HANDLES.

No. 0			No. 13	s.	d.
1	.	⎫	14	1	0
2	.	⎪	14	1	2
3	.	⎬ 4d. each.	15	1	4
4	.	⎪	16	1	6
5	.	⎪	17	1	8
6	.	⎭	18	1	10
7	.	6d.	19	2	0
8	.	6d.	20	2	3
9	.	8d.	21	2	6
10	.	8d.	22	2	9
11	.	10d.	23	3	0
12	.	10d.	24	3	6

FLAT HOG HAIR BRUSHES IN TIN,
FOR OIL PAINTING.

With long Thin Hair, being particularly adapted for Painting Fur and Hair.

Ditto ditto, shorter and thicker in the Hair, a suitable and most useful Brush in Landscape Painting.

THESE BRUSHES ARE MADE TO SIMILAR SIZES, AND ARE THE SAME PRICES AS THE ORDINARY HOG TOOLS.

BLACK FITCH BRUSHES IN TIN,
FOR OIL PAINTING.
POLISHED CEDAR HANDLES.
FLAT OR ROUND.

No. 1				s.	d.	No. 7				s.	d.
2	.	.	.	0	4	8	.	.	.	0	8
2	.	.	.	0	4	8	.	.	.	0	8
3	.	.	.	0	5	9	.	.	.	0	10
4	.	.	.	0	5	10	.	.	.	0	10
5	.	.	.	0	6	11	.	.	.	1	0
6	.	.	.	0	6	12	.	.	.	1	0

THESE BRUSHES ARE MADE THE SAME AS THE SABLES IN TIN.

FLAT SABLE BRUSHES IN TIN,
FOR OIL PAINTING.
POLISHED CEDAR HANDLES.

	s.	d.		s.	d.
No. 1	0	4	No. 7	1	0
2	0	5	8	1	3
3	0	6	9	1	6
4	0	7	10	1	9
5	0	8	11	2	0
6	0	9	12	2	3

ROUND SABLES IN TIN,
FOR OIL PAINTING.
WITH POLISHED CEDAR HANDLES.

	s.	d.		s.	d.
No. 0	0	4	No. 5	0	8
1	0	4	6	0	9
2	0	5	7	1	2
3	0	6	8	1	6
4	0	7	9	1	10

RED SABLE HAIR PENCILS,
FOR OIL PAINTING.

		s.	d.
Goose Quill	. . . .	0	8
Duck "	. . . .	0	6
Crow "	. . . .	0	4

SABLE WRITING PENCILS.

BADGER SOFTENERS.

		s.	d.			s.	d.
No. 1	each	0	6	No. 7	each	3	0
2	"	0	9	8	"	3	6
3	"	1	0	9	"	4	0
4	"	1	6	10	"	4	6
5	"	2	0	11	"	5	6
6	"	2	6	12		6	6

FLAT HOG HAIR VARNISHING BRUSHES.
WARRANTED TO RESIST TURPENTINE,
One Shilling per inch.

GROUND GLASS SLABS,
FOR GRINDING FINE COLOURS.

					s.	d.
Glass Slabs set in Mahogany Frames, 6 inches by 6	.				3	6
Ditto	ditto	8	„	8 .	5	0
Ditto	ditto	10	„	10 .	6	6
Glass Slabs only	. . . 6	„	6	. .	2	0
Ditto	. . . 8	„	8	. .	3	0
Ditto	. . . 10	„	10	. .	4	0

Glass Mullers, 1 inch diameter, 8d.——1¼ inch ditto, 10d.——
1½ inch ditto, 1s. 3d.——2 inch ditto, 1s. 6d.

BARTHOLOMEW'S EASEL FOR SKETCHING IN OIL.

INVENTED BY V. BARTHOLOMEW, ESQ., FLOWER PAINTER IN ORDINARY
TO HER MAJESTY AND H. R. H. THE DUCHESS OF KENT.

This Easel is equally adapted for the Studio and for out-door Sketching. It possesses the advantages of extreme portability, and of taking up very little space when open for use; and it is, therefore, well adapted for exhibiting and painting Pictures in Galleries or Studios. Price 15s.

PALETTES.

Spanish Mahogany Palettes, either oval or oblong, 8, 9, 10, 11,
and 12 inches . . . . . . per inch 2d.
Ditto ditto ditto 13 inches and upwards „ 2½d.
Ditto ditto ditto folding . . 2s 6d. to 3s. 6d.
Satin Wood, either oval or oblong, 8, 9, 10, 11, and
12 inches . . . . per inch 3d.
Ditto ditto ditto 13 inches to 15 inches „ 3½d.
Ditto ditto ditto 16 inches and upwards „ 4d.

DEAL AND MAHOGANY EASELS.

	s.	d.
Deal Closing Easel, 6 feet high	7	0
Mahogany ditto ditto	10	6
Deal folding Portable Easel	14	0
Mahogany ditto ditto	21	0
Deal Framed or Standing Easel	12	0
Mahogany ditto	16	0

MAHOGANY RACK EASELS, at 26s., 32s., 40s., & 50s.

MILLBOARD SKETCHING FRAME.

This Frame is made with a *rabbit* on each side, a millboard being placed upon both of them. On reversing the boards after painting, the wet sketches are brought face to face, without touching each other, and, at the same time, are perfectly protected.

PRICES OF FRAMES.

		s.	d.				s.	d.
12 inches by 9	.	2	3	18 inches by 12	.	.	3	6
14 „ 10	.	2	6	17 „ 13	.	.	3	6
15 „ 11	.	2	9	20 „ 14	.	.	4	0
16 „ 12	.	3	0					

ANY SIZE MADE TO ORDER ON A SHORT NOTICE.

MAHL STICKS.

		s.	d.
Bamboo Mahl Sticks		1	0
Ditto, long . . 1s. 6d. & 2		2	6
Jointed Mahl Sticks . . 2s. 6d., 3s., & 3		3	6
Telescope Mahl Sticks		4	0

TIN AND JAPANNED WARE FOR OIL PAINTING.

		s.	d.
Plain Tin Dippers or Palette Cups . . each		0	4
Ditto ditto double . . „		0	8
Japanned ditto ditto, various shapes . . „		0	6
Ditto ditto, double . . . „		1	0
Brush Washers, japanned . . . „		2	0
Ditto ditto, double . . . „		3	6
Brush or Smudge Pans . . . „		2	0

PALETTE KNIVES.

	s.	d.
Steel Palette Knives, with cocoa handles, various patterns and sizes . . . 6d., 7d., 8d., 9d., 10d., and	1	0
Ditto, with Ebony Handles . . . . . 9d. to	1	6
Ditto, with Ivory Handles . . . . 1s. to	2	0
Trowel Palette Knives, various sizes . from 1s. 6d. to	3	0
Ivory Palette Knives	1	0
Horn ditto	0	8

CHALKS, CHARCOAL, ETC.

		s.	d.
Sketching Charcoal	per doz.	0	6
Italian Chalk	per oz.	0	6
Pipe Clay in Sticks	per doz.	0	6
Permanent White Chalk, various degrees of hardness	per oz.	0	6
Red Chalk	,,	0	8
Soft French Stumping Chalk	per stick	0	1
Lithographic Chalk	per doz.	2	0
Conté Chalk, 1, 2, and 3	,,	0	6

GRECIAN PAINTING.

C. E. CLIFFORD MANUFACTURES ALL THE MATERIALS REQUIRED IN GRECIAN OR PERSIAN PAINTING, INCLUDING THE STUMPS, CHALKS, BRUSHES, POWDER COLOURS, ERASERS, PREPARED PAPER, CANVAS, &c., &c.

PREPARED PAPER FOR GRECIAN PAINTING, 21 INCHES BY 15, PER SHEET, 1s.

PREPARED CANVAS FOR GRECIAN PAINTING, ON STRETCHING FRAMES, TO ORDER.

ILLUSTRATED CATALOGUE

OF

APPARATUS & MATERIALS

USED IN THE ART OF

PHOTOGRAPHY,

MANUFACTURED AND PREPARED BY

C. E. CLIFFORD,

PHOTOGRAPHIC INSTRUMENT MAKER,

Operative Chemist,

AND ARTISTS' COLOURMAN,

30, PICCADILLY, LONDON, W.

C. E. CLIFFORD begs to submit to Amateurs and Artists the following Catalogue of Apparatus, Materials, and Chemicals, used in the practice of Photography; most of the Articles enumerated are manufactured under his own superintendence, he is therefore able to guarantee, with the greatest confidence, every article made and sold by him. His Cameras and Apparatus are manufactured from the best seasoned materials, and with every new improvement: his Lenses are unsurpassed by any makers, and the prices extremely moderate; he will undertake to guarantee or exchange any Lens of his own manufacture if not approved.

C. E. CLIFFORD respectfully informs Amateurs wishing to practise Photography, that he has entirely devoted his Photographic Rooms for the purposes of Instruction. Persons purchasing Apparatus at his Establishment will be taught the Art gratuitously until perfect.

C. E. CLIFFORD also begs to inform gentlemen about to proceed to India or the Colonies, that he has Apparatus especially adapted for foreign climates, and keeps constantly ready-fitted every description of Photographic Apparatus, from £8 to 50 Guineas the set.

C. E. CLIFFORD'S

Warranted Compound Achromatic Lenses for Portraits.

No.		£	s.	d.
1.	Compound Achromatic Lenses, mounted in handsome brass tubes, with rack and pinion adjustment, the lenses 1½ in. diameter, 3 in. focus, for stereoscopic and Carte de Visite pictures, fitted with disphragms	2	10	0
2.	Ditto, ditto, ¼ plate, the lenses 1½ in. diameter, 4½ in. focus, producing pictures, 4½ by 3½ and under.. ..	2	10	0
3.	Ditto ditto, the lenses 2½ in. diameter, 6½ in. focus, producing pictures 5 by 4 and under	4	4	0
4.	Ditto ditto, ½ plate, the lenses 2½ in. diameter, 7½ in. focus, producing pictures 6½ by 4½ and under.. ..	5	10	0
5.	Ditto ditto, whole plate, the lenses 3½ in. diameter, 10 in. focus, producing pictures 6½ by 6½ and under	12	12	0

These combinations are fitted with diaphragms, and are strongly recommended for giving extremely fine and correct definition both at the centre and margin of the picture, and for rapidity of action, unsurpassed.

Compound Achromatic Lenses of Foreign Manufacture.

	SIZES.	£	s.	d.		£	s.	d.
No. 1, for Portraits ..	4½ by 3½	0	16	0	......	1	5	0
2, ...	.. 6½ .. 4½	2	10	0	......	3	10	0
3, ...	.. 8½ .. 6½	6	10	0	......	8	8	0

C. E. CLIFFORD'S

Warranted Achromatic Landscape Lenses.

No.		£	s.	d.
1.	Achromatic Lens, mounted in handsome brass tubes, the lens 1⅛ in. diameter, for stereoscopic pictures ..	1	0	0
2.	Ditto ditto, with rack and pinion adjustment, the lens 1⅜ in. diameter, 7 in. focus, producing pictures 5 by 4	1	10	0
3.	Ditto ditto, the lens 2⅛ in. diameter, 10 in. focus, producing pictures 7 by 6	2	2	0
4.	Ditto ditto, the lens 2⅜ in. diameter, 12 in. focus, producing pictures 9 by 7 .. .. —	3	3	0
5.	Ditto ditto, the lens 3 in. diameter, 15 in. focus, producing pictures 10 by 8	5	0	0
6.	Ditto ditto, the lens 3⅛ in. diameter, 18 in. focus, producing pictures 12 by 10	6	10	0
7.	Ditto ditto, the lens 4 in. diameter, 20 in. focus, producing pictures 15 by 12	8	6	0
8.	Ditto ditto, the lens 4⅛ in. diameter, 22 in. focus, producing pictures 16 by 13	10	0	0
9.	Ditto ditto, the lens 5 in. diameter, 26 in. focus, producing pictures, 18 by 16	12	0	0

C. E. CLIFFORD'S

IMPROVED DOUBLE-BODY FOLDING CAMERA.

Is adapted for either Portraits or Landscapes, and can be shut up into a very portable form for travelling.

It is made of the best Spanish mahogany, French polished, with a Vertical and Horizontal Sliding Front for adjustment of Foreground and Sky, and may be had with or without the corners brass bound.

		SIZE.					£	s.	d.	With Brass Binding. £	s.	d.
No. 1.	For Pictures..	7 by	6	......			4	10	0	5	10	0
2.	....	.. 9 ..	7	......			5	5	0	6	6	0
3.	....	.. 10 ..	8	......			6	6	0	7	7	0
4.	....	.. 11 ..	9	......			7	10	0	8	10	0
5.	....	.. 12 .. 10		......			8	8	0	9	9	0
6.	....	.. 15 .. 12		......			11	5	0	13	0	0
7.	....	.. 16 .. 13		......			12	10	0	16	0	0
8.	....	.. 18 .. 16		......			17	10	0	21	0	0
9.	....	.. 22 .. 20		......			21	0	0	24	0	0
10.	....	.. 24 .. 22		......			23	10	0	27	0	
11.	....	.. 26 .. 24		......			28	0	0	30	0	0
12.	....	.. 33 .. 30		......			31	10	0	36	0	0

Folding and Rigid Cones for Doubling the Focal Length of the Lens for Copying Pictures adapted to any Camera.

COPYING CAMERAS FOR ENLARGING PICTURES.

SUPERIOR SLIDING-BODY CAMERAS.

Improved Sliding-body Cameras of the best Spanish Mahogany,
French polished, and of superior workmanship, the Camera and
Backs with or without Brass Binding.

		sizes.		£ s. d.	Brass Binding. £ s. d.
No. 1.	For Plates ..	5 by 4	......	1 14 0	2 10 0
2.	..	6½ .. 4½	......	2 8 0	3 8 0
3.	..	7 .. 6	......	3 0 0	4 5 0
4.	..	8½ .. 6½	......	4 3 0	5 10 0
5.	..	9 .. 7	......	5 0 0	6 5 0
6.	..	10 .. 8	......	5 15 0	7 5 0
7.	..	11 .. 9	......	7 0 0	8 10 0
8.	..	12 .. 10	......	8 0 0	10 5 0
9.	..	15 .. 12	......	10 0 0	13 10 0
10.	..	16 .. 13	......	13 0 6	16 0 0
11.	..	18 .. 16	......	17 10 0	21 0 0
12.	..	22 .. 20	......	22 10 0	26 10 0
13.	..	24 .. 20	......	27 0 0	31 10 0
14.	..	26 .. 24	......	33 0 0	37 10 0

SLIDING-BODY CAMERAS.

Sliding-Body Cameras, one body sliding in the other, of good Honduras Mahogany, French polished, with one single back and two Inner Frames for Collodion, Focus Screen, &c.

Horizontal or Vertical.						Square Cameras.						
No.				£	s.	d.					£	s.	d.
1.	For Plates 5 by 4	1	4	0	5 in. For Plates 4½ by 3½	1	6	0
2.	.... 6½ .. 4½	1	15	0	6 .... 5 .. 4	1	18	0
3.	.... 7 .. 6	2	5	0	7½ .... 6½ .. 4½	2	10	0
4.	.... 8½ .. 6½	3	0	0	9 .... 8 .. 6	3	15	0
5.	.... 9 .. 7	3	10	0	10 .... 8½ .. 6½	4	5	0
6.	.... 10 .. 8	4	5	0	11 .... 10 .. 8	5	5	0
7.	.. . 11 .. 9	5	0	0	12 .... 11 .. 9	6	15	0
8.	.... 12 .. 10	6	15	0	13 .... 12 .. 10	7	0	0
9.	.... 15 .. 12	9	0	0	16 .... 15 .. 12	9	15	0
10.	.... 16 .. 13	10	0	0	17 .... 16 .. 13	10	15	0
11.	.... 18 .. 16	11	10	0	20 .... 18 .. 16	15	0	0
12.	.... 22 .. 20	16	0	0	24 .... 22 .. 20	16	10	0
13.	.... 24 .. 20	19	0	0	26 .... 24 .. 22	24	0	0
14.	.... 26 .. 24	23	0	0	28 .... 26 .. 24	27	0	0

C. E. CLIFFORD's new Instantaneous Shutter adapted to any Camera.

STEREOSCOPIC CAMERAS.

Fig. 6.

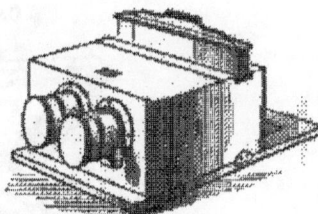

| Fig 7.

	£	s.	d.
LATIMER CLARKE'S Stereoscopic Camera arrangement on Parallel Laths, (Fig. 6)	3	3	0
Cameras adapted for 2 Lenses, for taking 2 Stereoscopic Pictures at one time (Fig. 7)	2	10	0
Extra Backs fitted to Latimer Clarke's Stereoscopic Cameras for Portraits, up to 5 inches by 4	0	15	0

IMPROVED DARK BOX. | PORTABLE BOX CAMERA.

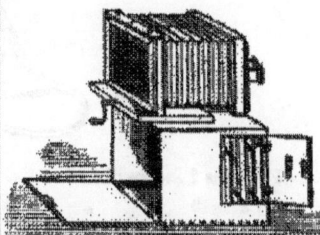

	£	s.	d.
Dark Box and Back for Stereoscopic Camera.	3	0	0
Ditto ditto, 9 by 7 ..	3	15	0
Ditto ditto, 10 by 8 ..	4	10	0

Dark Boxes Fitted to any Camera.

New Portable Binocular Stereoscopic Camera, with Bellows Body, Screw Movement for Focussing, 1 single and 3 double dark slides fitted in mahogany case—outside dimensions, 8½, 6 by 5½ inches; weight, 5 lbs.—Price, £5 10s.

SOLID LEATHER CASES, FOR FOLDING CAMERAS.

			£	s.	d.
For a Camera	7 by	6	1	10	0
....	9 ..	7	1	14	0
....	10 ..	8	1	16	0
....	11 ..	9	2	4	0
....	12 ..	10	2	12	0
....	15 ..	12	3	15	0

LEATHER SLING CASES, FOR LENSES.

			£	s.	d.
For Lens 2 inches diameter			0	10	0
„ 3 „ „			0	13	0
„ 4 „ „			0	16	0

CAMERAS IN WALNUT, of French Manufacture.

	VERTICAL			SQUARE.		
	£	s.	d.	£	s.	d.
Quarter Plate	0	10	0	0	14	0
Half ditto	0	18	0	1	3	0
Whole ditto	1	0	0	1	10	0

DOUBLE AND SINGLE BACKS.

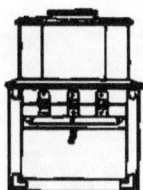  

Fig. 11.   Fig. 12.   Fig. 13.

Double Backs for Paper, of the best Spanish Mahogany, French
polished. (Fig. 12.) .

Single Backs with two Inner Frames for Collodion Process.
(Figs. 11 and 13).

No.		Size.	Double Backs. £ s. d.	Single Backs. £ s. d.	Brass bindg. ext. £ s. d.
1, for Pictures,	7 by 6	1 2 0 ..	1 0 0 ..	0 4 0
2,	....	8½ .. 6½ ..	1 5 0 ..	1 2 0 ..	0 4 0
3,	....	9 .. 7 ..	1 8 0 ..	1 4 0 ..	0 4 6
4,	....	10 .. 8 ..	1 12 0 ..	1 8 0 ..	0 5 0
5,	....	11 .. 9 ..	1 15 0 ..	1 10 0 ..	0 5 6
6,	....	12 .. 10 ..	2 0 0 ..	1 12 0 ..	0 5 6
7,	....	15 .. 12 ..	2 15 0 ..	2 5 0 ..	0 6 0
8,	....	16 .. 13 ..	3 0 0 ..	2 10 0 ..	0 6 0
9,	....	18 .. 16 ..	3 10 0 ..	2 15 0 ..	0 6 6
10,	....	22 .. 20 ..	4 0 0 ..	3 5 0 ..	0 7 0
11,	....	24 .. 22 ..	4 10 0 ..	4 0 0 ..	0 8 0
12,	....	26 .. 24 ..	5 5 0 ..	4 10 0 ..	0 9 0
13,	....	30 .. 26 ..	6 10 0 ..	6 0 0 ..	0 10 0

GUTTA-PERCHA DIPPING BATHS.
(Fig. 14.)

	s. d.		s. d.
Quarter-plate	2 0	to	3 6
Half ditto	3 6	„	5 6
Whole ditto	5 6	„	8 0

MOUNTED GUTTA-PERCHA BATHS.

Fig. 14.   Fig. 15.   Fig. 16.

Gutta-Percha Baths, in Mahogany Case, French polished, Watertight, for Plates of Sizes as below. (Fig. 15.)

No.	Size.	£	s.	d.
1	5 by 4	0	16	0
2	6¼ .. 4¾	1	0	0
3	9 .. 7	1	6	0
4	10 .. 8	1	10	0
5	11 .. 9	1	15	0
6	12 .. 10	2	0	0
7	13 .. 12	2	15	0
8	16 .. 13	3	0	0
9	18 .. 16	3	15	0

Glass Baths fitted as above, of the various sizes.

BRASS CLAMPED GUTTA-PERCHA BATHS.
(Fig. 16.)

	£	s.	d.
Quarter Plate	0	7	6
Half ditto	0	9	0
Whole ditto	0	12	0
Stereoscopic	0	9	0

PRESSURE FRAMES OF IMPROVED PATTERN.

 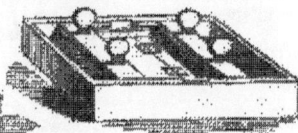

Fig. 17.   Fig. 18.

Pressure Frames in Oak, No. 17, or Mahogany, No. 18, with Hinged
Back, for viewing the Picture while printing.

	Size.	Oak.			Mahogany, polished.		
		£	s.	d.	£	s.	d.
For Pictures,	5 by 4	0	4	6	0	8	6
....	7 .. 6	0	7	0	0	13	6
....	9 .. 7	0	8	0	0	16	0
....	10 .. 8	0	10	0	0	17	6
....	11 .. 9	0	12	0	0	19	0
....	12 .. 10	0	15	0	1	1	0
....	13 .. 11	0	18	0	1	5	0
....	14 .. 12	1	0	0	1	10	0
....	16 .. 13	1	5	0	1	18	0
....	19 .. 17	1	10	0	2	5	0
....	23 .. 21	1	16	0	2	15	0
....	25 .. 23	2	2	0	3	5	0

HEAD RESTS.

	£	s.	d.
Head Rest, to attach to a chair	0	2	0
Ditto, jointed	0	5	0
Standard Head Rest, with Sliding Adjusting Tube for steadying the Head, either sitting or standing	3	3	0

BOXES FOR GLASS PLATES.

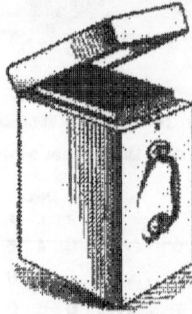

Boxes for holding one or two dozen Glass Plates, in walnut,
white wood, or mahogany, polished.

		Walnut.			White Wood.		Mahog. poled.	
	Size.	1 Doz.	2 Doz.	50.	1 Doz.	2 Doz.	1 Doz.	2 Doz.
		s. d.	s. d.	s. d.	s. d.	s. d.	s. d.	s. d.
For Plates,	3¼ by 2¼ ..	1 6	2 3	3 0	1 9	2 9	3 6	4 6
....	4½ .. 3½ ..	1 9	2 6	3 6	2 6	3 3	4 0	5 0
....	5 .. 4 ..	2 0	2 9	4 0	2 9	3 6	4 6	6 0
....	6¼ .. 4¾ ..	2 6	3 3	4 9	3 3	4 0	5 3	7 6
....	7 .. 6 ..	3 0	3 6	5 0	3 9	4 6	5 6	8 0
....	8¼ .. 6½ ..	3 6	4 0	5 6	4 0	5 0	6 3	8 6
....	9 .. 7 ..	4 0	5 0	6 0	4 3	5 6	7 0	9 0
....	10 .. 8 ..	5 0	6 0	8 6	4 9	6 0	8 0	10 0
....	11 .. 9 ..	....	....	....	5 6	7 0	9 6	12 0
....	12 .. 10 ..	....	....	....	6 3	7 9	10 6	13 0
....	13 .. 11 ..	....	....	....	7 0	9 0	12 0	15 0
....	15 .. 12 ..	....	....	....	8 0	11 0	16 0	20 0
....	16 .. 13 ..	....	....	....	9 0	12 0	17 0	22 0

Locks and Brass Handles to the above, from 2s. to 4s. extra.

If Brass-Screwed for India, from 1s. to 6s. extra.

Boxes to hold any number of Plates made to order.

CAMERA STANDS.

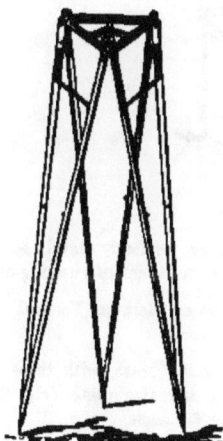 

Fig. 20.   Fig. 21.

Ash Tripod Stands, very firm and portable.

	£	s.	d.
Small Tripods .. .. from	0	5	0
½-in., of superior construction..	1	1	0
¾-in. ditto, with brass triangle top (Fig. 20)	1	5	0
1-in. ditto, ditto	1	8	0
Ditto ditto, very strong	1	15	0
Ash Tripod Stand, with jointed legs	1	16	0
Ditto ditto, with triangular top of mahogany, adapted			
for the support of large Cameras .. .. .. from	2	10	0
Table Stands, with Adjustments, for the Operating Room			
(Fig. 21) .. .. from	1	0	0
Ditto, very superior, in oak or mahogany, French polished,			
with screw to elevate or depress the table .. from £5 to	8	0	0

LEVELING STANDS.

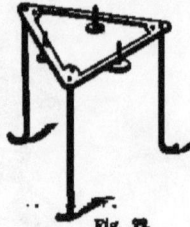

Fig. 22.

Leveling Stands for fixing glass plates perfectly level, so that solutions may be poured on to them without fear of running off.

Leveling Stand, with Iron Foot, Adjusting Screws, Vertical Rod and Rings, 7s. 6d. each.

Leveling Stand, triangular form, of bronzed brass, with three Feet and Adjusting Screws, for supporting any size plate, (Fig. 22) 4s. 6d., 6s. 6d., and 8s. 6d. each.

PORCELAIN PANS.

WITH FLAT BOTTOMS FOR WASHING AND PREPARING PAPER.

INCHES.		SHALLOW. £ s. d.	DEEP. £ s. d.
6 by 5		0 0 9	—
8 „ 6½		0 1 0	—
10 „ 7½		0 1 3	— 0 2 4
11 „ 9		0 1 6	— 0 2 8
13 „ 11		0 3 0	— 0 4 6
16 „ 11		0 4 0	— 0 6 0
18 „ 12½		0 5 6	— 0 8 0
19½ „ 14¼		0 13 0	— 0 16 0
20 „ 16		0 15 0	— 0 18 0
20 „ 18		0 18 0	— 1 0 0
22 „ 16		1 0 0	— 1 5 0
28 „ 22		1 10 0	— 1 15 0

Glass and Gutta-Percha Dishes at proportionate Prices.

IMPROVED PNEUMATIC PLATE HOLDER.

For securely holding Glass Plates of any size while coating with
Collodion, &c. .. .. 4s. 6d.

GLASS PLATES
FOR THE COLLODION PROCESS.

INCHES.	BEST PATENT PLATE, GROUND EDGES.			CROWN.	
		PER DOZ.		PER DOZ.	
		s. d.		s. d.	
2½ by 2	.. ..	1 0	—	0 4	
3½ „ 2⅝	.. ..	1 6	—	0 6	
4½ „ 3½	.. ..	2 0	—	0 9	
5 „ 4	.. ..	3 0	—	1 3	
6½ „ 4⅝	.. ..	4 4	—	2 0	
7 „ 6	.. ..	6 0	—	3 0	
8½ „ 6½	.. ..	8 0	—	4 0	
9 „ 7	.. ..	9 0	—		
10 „ 8	.. ..	12 6	—		
12 „ 10	.. ..	21 0	—		

GLASS GRADUATED MEASURES.

	s. d.			s. d.
1 ounce	1 0	10 ounce		4 0
2 „	1 4	16 „		4 6
4 „	2 0	20 „		5 0
6 „	3 0	32 „		6 6
8 „	3 6	Minums		1 0

Glass and Gutta-Percha Funnels from 4d. each.

Collodion Bottles, Graduated, 1 oz. 1s. 9d., 2 oz. 2s.

C. E. CLIFFORD'S PHOTOGRAPHIC TENT.

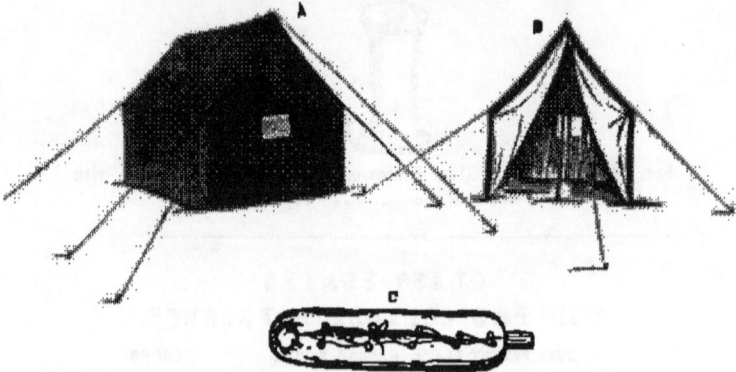

A. The Tent when Erected, 6 feet 6 high, containing 20 square feet.
B. The ditto, showing the Interior.
C. The same Folded for Travelling, 24 inches by 7 in diameter.

Price £4. 4s.

C. E. CLIFFORD would call particular attention to his Dark Tent, which is the most convenient and portable yet invented. It allows ample room for working the largest plates, is waterproof, stands well against wind, and can easily be put up by one person.

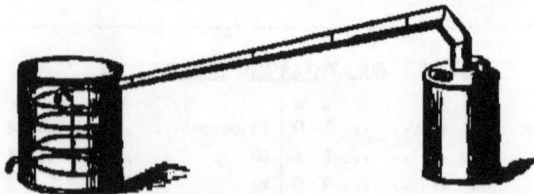

PORTABLE STILLS.

Still to hold One Gallon, with Worm Tub complete, for use over a common fire, £1. 1s.; Packed in Box for Travelling, £1. 5s.

B

SCALES AND WEIGHTS.

Fig. 27.

			s.	d.
Complete, in Oak Box, with Metal Pans	..	.. .. ..	2	6
Ditto	ditto	Glass ditto	5	0
Ditto	Mahogany Box, ditto	.. ..	7	6
Ditto	ditto	ditto superior..	12	0
Ditto	ditto, with Brass Pillar to screw as standard on			

top of Box, Glass shifting Pan, with Drawer, &c. Fig. No. 27,
From £1. 15s. to £2. 2s,

MISCELLANEOUS.

		s.	d.	
Spirit Levels	.. .. from	4	0	each.
Ditto Lamps	.. .. „	2	0	„
Horn and Wood Tongs	.. ..	1	0	⅌ pair.
Photographic Pins for Suspending Paper	.. ..	1	0	⅌ box.
Brushes for spreading Solutions	.. ..	1	0	each.
Plate Cleaners, with Screw to fit any size Plate	..	5	6	„
American Clips, for suspending paper	.. .. ..	1	0	⅌ doz.

Marine Glue for Cementing Glass.

Retort and Filtering Stands.   Gold and Silver Shells.

Diamond Files for Glass Plate, each 1s. and 1s. 6d.

Metal Tablets, for Positive Pictures.   Test Paper.

STEREOSCOPES.

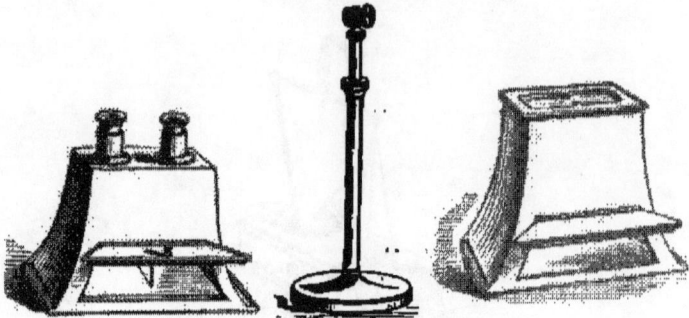

			s.	d.
Stereoscopes, covered with black morocco paper · ·· · ··	from	2	0
Ditto	in mahogany ·· ·· ·· ·· ·· ·· ·· ··	„	8	0
Ditto	in polished mahogany or walnut, with highly finished eye pieces, hinged flap, of the best construction ·· ·· ·· ·· ··	12s. to £1	1	0
Stereoscopic Stand ·· ·· ·· ·· ·· ·· ·· ··	from	15	0

STEREOSCOPIC PICTURES.

Stereoscopic Views in great variety from all parts of the globe, the Crystal Palace, &c., on Glass, Plate, and Paper.

Stereoscopic Pictures, on Paper, from 9s. per dozen.

Stereoscopic Pictures, of superior character, having finest detail, and taken by superior Artists, 2a., 2s. 0d. to 10s. 6d.

ALBUMS AND GUARD BOOKS.

C. E. CLIFFORD has always in Stock a great variety of Albums for Mounting Photographic Pictures, of every size and style of Binding.

Photographic Albums made to Order, and Photographs Mounted.

B 2

PHOTOGRAPHIC PAPERS.

	Size.	Negative. per Qr. s. d.	Positive. per Qr. s. d.
Canson Freres'	22½ by 17½ ..	3 0 ..	4 0
Whatman's	19 .. 15 ..	3 0 ..	3 0
Turner's	18 .. 15 ..	5 0 ..	3 0

NEGATIVE PAPER.

	Size.	Per Quire. £ s. d.
Waxed Paper	17½ by 11½ ..	0 9 0
Ditto and Iodised	ditto ..	0 13 0

EXTRA SENSITIVE FINE NEGATIVE PAPER.

	Size	£ s. d.
Plain	22½ by 17½ ..	0 14 0
Waxed	ditto ..	1 10 0
Ditto and Iodised	ditto ..	2 5 0

SALTED PAPERS,

PREPARED AND READY FOR EXCITING.

	Size	£ s. d.
Salted Chloride Barium ..	29½ by 17½ ..	0 4 6
Ditto Ditto Sodium ..	ditto ..	0 4 6
Ditto Ditto Ammonium	ditto ..	0 4 6

ALBUMINISED SAXE PAPER.

C. E. CLIFFORD's Extra Superfine Saxe Albuminised Paper,
22½ by 17½ .. 12s. per Quire.

This Paper can be most highly recommended for producing pictures
of the finest detail and exquisite tone.

		£ s. d.
Albuminised	22½ by 17½ ..	0 5 0
Ditto Extra	ditto ..	0 6 0
Ditto Super Extra	ditto ..	0 12 0
White Bibulous Paper	1s. &	0 1 6

Paper Rounds, for Filtering, from 6in. to 20in. in diameter,
from 9d. the Hundred.

Yellow Paper for Laboratory, 3s. per Quire.

SOLID LEATHER PHOTOGRAPHIC CHEMICAL BOX.

CONTAINING A COMPLETE SET OF CHEMICALS, FOR TRAVELLING.

Size, 9¼ inches by 8, and 6 inches deep.

C. E. CLIFFORD'S

Chemically Prepared Photographic Colours.

These Colours having been prepared with the greatest care and purity by a chemical process, will be found admirably adapted for Tinting the Glass and Daguerreotype Plates; also Portraits and Landscapes taken on Paper. The Colours are all strictly permanent, a desideratum hitherto unattained; and from their great variety of Tint, effects can be obtained with the greatest ease and brilliancy, even by persons previously unacquainted with Colouring.

Photographic Colours, in mahogany box, containing an assortment of Twelve Chemically Prepared Colours in small bottles, an assortment of Brushes, Gold and Silver Shells, 9s.

Separate Colours, in bottles .. 6d. each.
Camel Hair Brushes .. .. 1s. per dozen.
Sable ditto .. .. 3s. „

PURE CHEMICAL PREPARATIONS, &c.

REQUISITE IN THE VARIOUS PROCESSES OF THE PHOTOGRAPHIC ART.

		s.	d.
ACIDS.			
Acetic Glacial	per ounce	0	6
Ditto Crystallized	"	1	0
Formic	"	0	4
Gallic	"	1	6
Pyro-Gallic (pure) White Crystals	"	8	0
Hydrochloric	per pound	1	6
Nitric (pure)	"	1	6
Citric	per ounce	0	6
Sulphuric	per pound	1	6
Tannic	per ounce	1	6
AMMONIA, Concentrated	"	0	9
AMMONIUM.			
Bromide	"	4	6
Chloride	"	3	0
Iodide	"	3	0
BARIUM, Chloride	"	0	9
BARYTA, Nitrate	"	0	3
BENZOLE (pure)	per pound	3	0
BROMINE (pure)	per ounce	3	0
CADMIUM.			
Bromide	"	4	6
Iodide	"	4	0
CALCIUM.			
Iodide	"	4	6
Bromide	"	4	0
CHARCOAL, Animal (pure)	"	0	6
COLLODION.			
Plain	per pound	10	6
Iodized	"	10	6
Iodizing Solution	"	10	6
COTTON, Fine Carded	per ounce	0	3
ETHER, Sulphuric, for Collodion	"	0	6
GLYCERINE (pure)	"	0	9
GOLD.			
Chloride, unadulterated with Soda	per 15 grain bottle	2	9
Hyposulphite (Sel d'Or)	"	3	0
IODINE.			
Re-sublimed	per ounce	2	0
Bromide (Sol.)	"	4	0
Chloride	"	4	0
Tincture, prepared for Wax Paper	"	0	8

		s. d.
IRON.		
Ammonia-citrate	per ounce	1 0
Iodide	„	0 3
Protosulphate (pure)	per pound	0 10
KAOLIN	per ounce	0 2
LIME, Bromide	per bottle	3 0
LEAD.		
Nitrate	per ounce	0 2
Acetate	„	0 2
MERCURY, Bi-chloride	„	0 6
MILK, Sugar of	„	0 3
OXYMEL	„	0 4
POTASSA, Nitrate (pure)	„	0 3
POTASSIUM.		
Bromide	„	3 0
Cyanide	„	0 3
„ (pure)	„	0 4
Fluoride	„	2 0
Iodide	„	2 0
Double Iodide for preparing Talbotype Paper	„	2 6
ROUGE, finely prepared	„	0 6
SILVER.		
Chloride	„	8 0
Iodide	„	8 0
Nitrate, Crystallized	„	4 0
„ Re-Crystallized	„	4 6
„ Fused	„	5 0
Solution, Prepared, for the Collodion Bath	per pint	6 6
SODA, Hypo-sulphite	per pound	0 10
Acetate	per ounce	0 3
SODIUM.		
Chloride (pure)	„	0 2
Fluoride	„	1 0
TONING, or Hypo-Colouring Bath	in pint bottles	6 0
TRIPOLI	per ounce	0 6
VARNISH.		
Amber and Chloroform	„	1 0
White Crystal	per bottle	0 6
Jet for Backing	„	0 6

The above Prices are subject to variation.

All new Preparations made and supplied immediately they
are introduced.

AGENT FOR PONTING'S COLLODION.

MOROCCO CASES.

Superior Real Morocco Cabinet Cases of London make, in various sizes.

French Oval Velvet Cases, for Miniatures, from 2s. 6d.

Cases of every description made to Order.

ENGLISH MOROCCO CASES.

	2½ × 2	3¼ × 2¾	4¼ × 3¼	5 × 4
Morocco Cases, Mats and Glasses	3 4	5 6	9 0	18 0
Ditto, Ditto, Gilt inside	4 0	6 6	10 6	20 0
Ditto, Ditto, Extra	4 0	7 0	10 0	24 0
Ditto, Ditto, Ditto, Gilt inside ...	5 0	8 0	11 6	26 0
Ditto, Ditto, Silk Velvet Cushions, Mats and Glasses ...	5 0	8 6	12 6	18 0
Ditto, Ditto, Ditto, Gilt inside ...	5 6	9 6	14 0	23 0
Ditto, Ditto, Ditto, Ditto	7 0	10 6	14 0	26 0
Ditto, Ditto, Ditto, Ditto, gilt inside	8 0	12 6	17 0	20 0
Ditto Trays, Gilt inside, Mats and Glasses	2 8	3 6	5 6	12 0
Ditto, Ditto, Silk Ditto Ditto ...	4 0	0 0	8 0	12 0

AMERICAN MOROCCO CASES.

Morocco Cases, Mats, and Glasses	6 0	8 0	12 6
Ditto, Ditto, Gilt inside	6 6	8 9	13 6
Ditto, Ditto, Silk Velvet	7 3	11 3	14 6
Ditto, Ditto, Gilt inside	7 9	12 0	15 6

MATS AND PRESERVERS.

	2¼x2	3¼x2¼	4¼x3¼	5x4	6¼x4½	8¼x6¼
Mats, Oval, Cushion, or Dome Shape	9 6	0 10	1 4	3 0		
Chased Ditto, Ditto	0 9	1 8	2 0			
German Ditto, Ditto ...	2 6	5 0	6 6			
American Preservers ...	0 6	0 8	1 0	2 6		

PASSEPARTOUTS,
OVAL, CUSHION, OR DOME SHAPES.

Black, Brown, or Tortoiseshell Ground	1 8	2 2	2 10	5 0	6 6	12 0
Ditto, Ditto, Porcelain Bevel	3 0	3 6	4 0	8 0	10 0	12 0
White Ground Gold Ditto	2 6	3 0	4 0	6 9	9 6	18 0
Gold Ditto	7 0	8 0	9 0	15 0	19 6	40 0
White Ditto, Broad Margins	4 6	5 6	6 6	8 6	9 6	16 0
Super. Bristol Ditto... ...	8 0	10 0	12 0	15 0	21 0	30 0

STEREOSCOPIC PASSEPARTOUTS,
From 3s. per Dozen.

FRAMES FOR PASSEPARTOUTS.

Renaissance Pattern, Black or Rosewood	4 6	5 0	5 6	7 0	8 6	14 0
Old Oak, richly Ornamented, Oval or Cushion	9 0	10 0	12 0	15 0	21 0	36 0
Plain Gilt	5 0	6 0	7 0	8 0	10 0	18 0
Ditto with Corners	12 0	15 0	18 0	21 0	24 0	30 0
Gold, richly Ornamented, Oval or Cushion ... each	2 6	3 0	3 6	4 0	4 6	5 6

ESTIMATES OF
COMPLETE SETS OF PHOTOGRAPHIC APPARATUS.

FOR PORTRAITS.

No. 1.—Complete Set of Photographic Apparatus for Portraits 4½ in. by 3½ in. and under, comprising walnut expanding camera, fitted with double combination lens, rack and pinion adjustment, tripod stand, box of scales and weights, porcelain washing dish, gutta percha bath and dipper, graduated measures and funnel, filtering paper, one dozen glass, and bottles containing all the necessary chemicals, &c., packed in box £3 0 0

No. 2.—Complete Set of Photographic Apparatus for Portraits 4½ in. by 3½ in. and under, comprising Spanish mahogany expanding camera, French polished, fitted with double combination lens, rack and pinion adjustment, of C. E. CLIFFORD's own manufacture, tripod stand, box of scales and weights, porcelain washing dishes, gutta percha bath and dipper, graduated measures and funnel, filtering paper, glass plates and boxes, pressure frame, glass rod, albuminised paper, developing stand, and all the necessary chemicals in stoppered bottles for the positive and negative processes, packed in box with lock and key .. £7 7 0

No. 3.—Complete Set of Photographic Apparatus for Portraits 6½ in. by 4½ in. and under, comprising walnut expanding camera, fitted with double combination lens, rack and pinion adjustment, tripod stand, box of scales and weights, porcelain washing dishes, gutta percha bath and dipper, graduated measures and funnel, filtering paper, glass plates, and bottles containing all the necessary chemicals, &c., packed in box £6 6 0

No. 4.—Complete Set of Photographic Apparatus for Portraits 6¼ in. by 4¾ in. and under, comprising Spanish mahogany expanding camera, French polished, fitted with double combination lens, rack and pinion adjustment, of C. E. CLIFFORD's own manufacture, tripod stand, box of scales and weights, porcelain washing dishes, gutta percha bath and dipper, graduated measures and funnels, filtering paper, glass plates of three sizes and boxes, printing frame, glass rod, albuminised paper, developing stand, horn tongs, suspending pins, and all the necessary chemicals in stoppered bottles for the negative and positive processes, packed in box with lock and key £12 12 0

No. 5.—Complete Set of Photographic Apparatus for Portraits 8¼ in. by 6¼ in. and under, comprising walnut expanding camera, fitted with double combination lens, rack and pinion adjustment, tripod stand, box of scales and weights, porcelain washing dishes, gutta percha bath and dipper, graduated measures and funnel, filtering paper, glass plates and boxes, and bottles containing all the necessary chemicals, &c., packed in box £11 11 0

No. 6.—Complete Set of Photographic Apparatus for Portraits 8¼ in. by 6¼ in. and under, comprising Spanish mahogany expanding camera, French polished, fitted with double combination lens, rack and pinion adjustment, of C. E. CLIFFORD's own manufacture, tripod stand, box of scales and weights, porcelain washing dishes, gutta percha bath and dipper, graduated measures and funnels, filtering paper, glass plates of three sizes and boxes, printing press, glass rod, albuminised paper, developing stand, horn tongs, suspending pins, and large supply of all the necessary chemicals in stoppered bottles for the positive and negative processes, packed in box with lock and key .. £24 0 0

FOR STEREOSCOPIC VIEWS AND PORTRAITS.

No. 7.—Complete Set of Photographic Apparatus for Stereoscopic Views, comprising Spanish mahogany camera of the best construction, with adjusting screw on Latimer Clarke's principle, fitted with view lens, handsomely mounted, rack and pinion adjustment, of C. E. CLIFFORD's own manufacture, tripod stand, box of scales and weights, porcelain washing dishes, gutta percha bath and dipper, graduated measures and funnel, filtering paper, glass plates with ground edges in box, pressure frame, albuminised paper, developing stand, horn tongs, suspending pins, and all the necessary chemicals in stoppered bottles, packed in box with lock and key £9 9 0

No. 8.—Complete Set of Photographic Apparatus for Stereoscopic Portraits and Views, comprising Spanish mahogany camera of the best construction, with adjusting screw of Latimer Clarke's principle, fitted with double achromatic lens for portraits and extra lens for views, handsomely mounted, rack and pinion adjustment, of C. E. CLIFFORD's own manufacture, tripod stand, box of scales and weights, porcelain washing dishes, gutta percha bath and dipper, graduated measures and funnel, filtering paper, glass plates with ground edges in box, printing frame, albuminised paper, developing stand, horn tongs, suspending pins, and all the necessary chemicals in stoppered bottles, packed in box with lock and key £11 11 0

No. 9.—Complete Set of Photographic Apparatus for Stereoscopic Views, comprising Spanish mahogany camera of the best construction, fitted with two single achromatic lenses for views, with rack and pinion adjustment, of C. E. CLIFFORD's own manufacture, for taking the two views at the same instant, tripod stand, box of scales and weights, porcelain dishes, gutta percha bath and dipper, graduated measures and funnels, photographic papers, developing stand, horn tongs, suspending pins, and all the necessary chemicals in stoppered bottles, packed in box with lock and key £10 10 0

No. 10.—Complete Set of Photographic Apparatus for Stereoscopic Portraits and Views, comprising Spanish mahogany camera of the best construction, fitted with two double combination lenses for portraits, and two single adjusting lenses for views, handsomely mounted rack and pinion adjustment, of C. E. CLIFFORD's own manufacture, tripod stand, box of scales and weights, porcelain dishes, gutta percha bath and dipper, graduated measures and funnels, filtering paper, glass plates with ground edges in box, printing frame, albuminized paper, developing stand, horn tongs, suspending pins, and all the necessary chemicals in stoppered bottles, packed in box with lock and key .. £14 14 0

FOR VIEWS.

No. 11.—Complete set of Photographic Apparatus for Views, 6 in. by 5 and under, comprising mahogany expanding camera, French polished, fitted with single achromatic lens, rack and pinion adjustment, of C. E. CLIFFORD's own manufacture, tripod stand, box of scales and weights, porcelain dishes, gutta percha bath and dipper, graduated measures and funnel, filtering paper, glass plates in boxes, pressure frame, photographic papers, developing stand, and all the necessary chemicals in stoppered bottles, packed in box, with lock and key £10 0 0

No. 12.—Complete Set of Photographic Apparatus for Views, 7 in. by 6 and under, comprising mahogany expanding camera, French polished, fitted with single achromatic lens, rack and pinion adjustment, of C. E. CLIFFORD's own manufacture, tripod stand, box of scales and weights, porcelain dishes, gutta percha bath and dipper, graduated measures and funnel, glass plates in boxes, filtering paper, albuminized paper, printing frame, developing stand, horn tongs, suspending pins, and all the necessary chemicals in stoppered bottles, packed in box, with lock and key £13 0 0

No. 13.—Complete Set of Photographic Apparatus for Views, 9 in. by 7 and under, comprising mahogany expanding camera, French polished, fitted with single achromatic lens, rack and pinion adjustment, of C. E. CLIFFORD's own manufacture, tripod stand, box of scales and weights, porcelain dishes, gutta percha bath and dipper, graduated measures and funnels, glass plates in boxes, filtering paper, albuminized paper, printing frame, developing stand, horn tongs, suspending pins, and all the necessary chemicals in stoppered bottles, packed in box, with lock and key £15 0 0

No. 14.—Complete Set of Photographic Apparatus for Views 9 in. by 7 and under, comprising Spanish mahogany camera, with horizontal and vertical sliding front, for adjustment of sky and fore-ground, of best manufacture, French polished, fitted with single achromatic lens, handsomely mounted, rack and pinion adjustment, of C. E. CLIFFORD's own manufacture, ash tripod stand with metal top, box of scales and weights, porcelain dishes, gutta percha bath mounted in mahogany case, French polished, with screw top for securely carrying the solution while travelling, &c., graduated measures and funnels, patent plate glass in boxes, filtering paper, albuminized paper, printing frame, pneumatic holder, developing stand, horn tongs, suspending pins, glass rod, filtering stand, and a full supply of all the necessary chemicals, &c., in stoppered bottles, packed in iron-bound box, with lock and key.. .. .. £20 0 0

No. 15.—Complete Set of Photographic Apparatus for Views, 9 in. by 7 and under, comprising Spanish mahogany folding camera, with vertical and horizontal sliding front for adjustment of sky and fore-ground, of best manufacture, French polished, packed in solid leather case lined with baize for travelling, fitted with single achromatic lens handsomely mounted, rack and pinion adjustment, of C. E. CLIFFORD's own manufacture, with solid leather sling case for carrying the lens, ash tripod stand French polished with metal top, box of scales and weights, porcelain dishes, gutta percha bath mounted in mahogany case French polished, with screw top for securely carrying the solution while travelling, &c., graduated measures and funnels, patent plate glass with ground edges, packed in boxes, filtering paper, albuminized paper, mahogany pressure frame French polished, developing stand, horn tongs, suspending pins, glass rod, spirit level, pneumatic holder, filtering stand, and a full supply of all the necessary chemicals in stoppered bottles, packed in iron-bound box, with lock and key £25 0 0

No. 16.—Complete Set of Photographic Apparatus, for Views, 10 in. by 8 and under, with Spanish mahogany camera, of a similar description and fittings to that described in No. 14 Set, packed in iron-bound box, with lock & key.. £25 0 0

No. 17.—Complete Set of Photographic Apparatus for Views, 10 in. by 8 and under, with Spanish mahogany folding camera and case, of a similar description and fittings to that described in No. 15 Set, Packed in iron-bound box, with lock and key £30 0 0

No. 18.—Complete Set of Photographic Apparatus for Views, 11 in. by 9 and under, with Spanish mahogany camera, of a similar description and fittings to that described in No. 14 Set, packed in iron-bound box, with lock and key £30 0 0

No. 19.—Complete Set of Photographic Apparatus for Views, 11 in. by 9 in. and under, with Spanish mahogany folding camera and case, of a similar description and fittings to that described in No. 15 Set, packed in iron-bound box, with lock and key £42 0 0

No. 20.—Complete Set of Photographic Apparatus for Views, 12 in. by 10 in. and under, with Spanish mahogany camera, of a similar description and fittings to that described in Set No. 14, packed in iron-bound box, with lock and key £42 0 0

No. 21.—Complete Set of Photographic Apparatus for Views, 12 in. by 10 in. and under, with Spanish mahogany folding camera, of a similar description and fittings to that described in Set No. 15, packed in iron-bound box, with lock and key £50 0 0

No. 22.—Complete Set of Photographic Apparatus for Views, 15 in. by 12 in. and under, with Spanish mahogany camera, of a similar discription and fittings to that described in Set No. 14, packed in iron-bound box, with lock and key £55 0 0

Any of the above Cameras can be fitted with Double Achromatic Lens, for Portraits.

www.ingramcontent.com/pod-product-compliance
Lightning Source LLC
Chambersburg PA
CBHW021822190326
41518CB00007B/699